Proud to Be:
Writing by
American Warriors
Volume 7

Southeast Missouri State University Press • 2018

Proud to Be:
Writing by
American Warriors
Volume 7

Edited by James Brubaker

Partners in the Military-Service Literature Series

Proud to Be: Writing by American Warriors, Volume 7
Copyright by Southeast Missouri State University Press.
All rights reserved. Permission to reprint a particular
author's individual work will be granted upon that
author's request to the University Press.

ISBN: 978-1-7320399-0-2

First Published in the United States of America, 2018
Southeast Missouri State University Press
One University Plaza, MS 2650
Cape Girardeau, MO 63701
http://www.semopress.com

Cover photograph: "Patrolling the Arghandab" by Breanne M. Pye
Cover design: Carrie M. Walker

Southeast Missouri State University Press, founded in 2001, serves
as a first-rate publisher in the region and produces books, *Big Muddy:
Journal of the Mississippi River Valley*, *The Cape Rock* poetry journal, and
the Faulkner Conference series.

The Missouri Humanities Council is a 501(c)3 non-profit organization that was created in 1971 under authorizing legislation from the
U.S. Congress to serve as one of the 56 state and territorial humanities councils that are affiliated with the National Endowment for the
Humanities.

Contents

Fiction

Additional Writing and Photography

Essays

Poetry

Photography

Interviews

Foreword

It's infuriating, isn't it, not finding the right words? Sometimes, I have the hardest time even starting a sentence; it drives me nuts! I feel this way especially about my experiences as an Active Duty servicemember, U.S. Army, 1998-2004.

You see: I've always been a storyteller. Born and raised in Puerto Rico, I started babbling in Spanglish well before I was supposed to. When I was able to write out my name for the first time in Kindergarten, I arrived home with a feeling I now recognize as elation. I demanded my mother's full attention as I described how a cream-colored, rectangular sheet of paper now held the shaky skeleton of my newfound identity, all written with a ginormous #2 pencil on intermittent green lines. It wasn't until I was certain my mother heard every word I'd said that I showed her the evidence. I remember us both smiling over this unexpected treasure.

Growing up, I was the one everyone—and I do mean everyone—at school asked to write their last-minute essays. I made a profit from my talents, mostly in the form of candy, trendy lip-gloss, and math homework. My desperate (and lazy) clients were (mostly) satisfied with their B+'s and A-'s. I take a bit of pride in my stellar customer service back in those mischievous middle- and high school days; admittedly, I always kept the A+ stuff for myself.

Then, life exploded as it tends to do. My mother fell gravely ill: she had stroke after debilitating stroke. Through some research, I found out if I joined the Armed Forces I could try to make her my dependent and, hopefully, help her have some semblance of an enhanced quality of life.

I was inspired by the Army's energetic "Be All That You Can Be" slogan. I didn't know what I wanted to be, but the ads showed smiling and determined soldiers sharing important tasks in co-ed environments. My chest swelled with patriotic and feminist pride.

I was swayed by a compelling promise of a new frontier, one in which I could find a path in life while simultaneously helping my ailing mother. I committed my signature to the cause, wrote it down with fear—miles away from the bliss I'd first felt discovering what my name looked like when it's committed to paper. Little did I know that a life of stark contrasts would begin after I signed the proverbial dotted line.

I don't quite remember how, but I survived all the initial training. Looking back on it, parts of Basic Training at Ft. Jackson, South

Carolina, were actually quite fun. The child in me will always love to climb over unlikely obstacles. After several months of Advanced Individual Training, I was kept stateside for my first duty station. I lost myself in my job as an Intelligence Analyst, Report Writer, and Linguist. I worked hard: doggedly, obsessively hard. Sergeants, officers, and supervisors started to—quite literally—yell at me: "Go Home!"

I was determined to find that promised path, define the boundaries of what I believed was forthright, and find new ways to help my ailing mother. I was hyper-aware of the weight of adult responsibility I carried on my young shoulders. The burden often felt excessive, stifling.

One day my mother asked a crucial question.

"¿Que estás escribiendo?"

I couldn't quite answer her. I'd been writing intelligence reports like they were about to go gangbusters. Yet, I couldn't remember the last time I'd sent her a letter that also contained an absurd story, a crass limerick, or a silly haiku. I couldn't remember the last time I'd doodled on stories about damsels in distress who were actually highway robbers, or extraterrestrials masquerading as mediocre magician acts in Reno, NV. I couldn't remember the last time I'd written for pleasure. I'd forgotten about my words, my precious words.

I'm sure that creative amnesia trickled in long before the Army during my mother's sudden illness. One day, I was writing someone else's essay for sweets; the next, I had to start a life that included changing my bed-ridden mother's diapers. It's a sequence of events that will force you to become an adult fast, like it or not.

Later, as time passed while I served, I had to consciously shut down parts of me. Part of it was a side effect of navigating life as a cublicled soldier in a post 9-11 world. Mostly though, I couldn't write because of the other challenges, those that are still part of the package deal too many workingwomen have been handed since the beginning of time.

We've heard the term a lot lately, haven't we: Post-Traumatic Stress Disorder (PTSD). I've had it since my mother's illness, but it was exacerbated after I was sexually assaulted in the military by a higher-raking service member. Soon after, I was silenced in eerily effective ways any time I tried to report the crime. Out-of-touch leaders assured me my attacker's reputation and rank mattered more than the justice owed me. Many who heard of what happened chose to take his side. Perhaps this is due to our society's predilection for victim blaming. Perhaps denial was the fastest path towards regaining a sense of normalcy in the unit. Perhaps it was that few knew how to properly talk about these things. Acknowledgement and corroboration were probably too awkward an order. Throughout the process, I found myself quite isolated as my words were slighted, ridiculed, and ignored.

I internalized this culture of silence. Words became the enemy. Writing for leisure meant having to stop and think. It meant acknowledging and feeling. It meant making room for justified reactions. Those choices were simply not allowed at the time for me. Consequently, numbing—one of the stages of PTSD—became essential in my emotional and psychological survival. The only way to channel the righteous rage compartmentalized within was to pour it into my job. I marched on this way, in the manner many soldiers have.

Looking back at my time in service, I see that life lived in contrasts that manifested in extraordinary ways. The unwavering pride in wearing the uniform remained, but the respect for rank became superficial after the attack. I continued to dazzle supervisors with my steadfast work ethic, but the pride in a job well done faded and turned into tasteless drudge. Words were anathema and silence became a bizarre ally.

Soon after I was honorably discharged, I was reading a book about great quotes from great leaders to ready myself for a Women Veterans' networking event. "The truth is incontrovertible," attributed to Winston Churchill, stuck out like a mischievous gopher at a golf club. I chewed on this wisdom for a bit. Then a simultaneous mix of thought and feeling crept in: just like truth, words will find a way to emerge. I shook my head, of course. I knocked down that gopher of an idea and busied myself with preparations for the event.

The next day, after some mild networking, a fellow Veteran abruptly asked did "it" happen to me while I was serving.

"It?" I was startled by her openness.

"You know..."

I did know. Of course I knew. I didn't know how to talk about it though. I didn't know we *could* talk about it. Was that even allowed?

I couldn't answer her. I couldn't find the words. I'd never allowed myself to seek the words. After years and years and years, "It" remained inside a hastily lacquered box that'd been locked with ache and shame and silence.

She saw the conflict inside me. I thank God for her kindness and openness; she took the initiative and chose to speak for us both. We left the event with a third Veteran who overheard her. Over many cups of watered down, over-priced coffee, they both relayed their pain, their struggle, their healing process. I was captivated as they stumbled on the words, ambled towards the shape of their stories, and emerged with narratives that mirrored mine.

We became jittery from the acknowledgment, corroboration, and caffeine. We hugged, and cried, and laughed, all while wearing our best suits at a coffee shop filled with strangers.

We connected, in the only way honest words can connect humans.

Soon after, while I brewed my own, better-tasting coffee at home, I wondered about the shape of my own narrative. I sat down to write about "it." The words didn't come easily, not at first. Armed with the discipline serving had instilled in me—one of the many fine qualities learned while I served—I sat at my desk until some things were committed onto the lineless pages of a plain journal. I didn't stop until my coffee turned cold.

I read my words. My handwriting was shaky, just like it was when I first wrote my name down. There was also a viscous terror, deeper and more primal than what I felt the day I signed on to serve our country.

Slowly, recognition emerged. I was there. So were the pain, confusion, and feeling that the silence of many—including myself—betrayed me. Yet, I also saw strength and perseverance. I saw patience. I saw courage and care so strong they carried me through all the unwarranted, unconsented agonies society has imposed on me simply because I was born a woman—and a woman of color at that.

It took a while, but I started writing again. First, I sent my mother light comic strips about "The Adventures of Curly Hair Cosmic Queen vs. The Humidity Demon!" or "Wheelchair Mami Rolls Over The Screeching Rooster!" (No, you can't see these, because I can't draw. And they are terrible).

Anyway, my journaling took a life of its own. I started writing down my thoughts every day. Later, much, much later, after much needed time, therapy, and endless rounds of eye- and heart-opening conversations with many Veteran sisters and brothers over endless gallons of terrible coffee, I pursued a Master of Fine Arts in what I've always loved, what was there since the beginning: Creative Writing.

Even though she's still bed-ridden and has had numerous other battles with her health, my mother refuses to kick the bucket. (I'm certain she has traces of Galápagos turtle in her DNA. Perhaps we are the descendants of Methuselah). By now, I've also gained a pretty good idea about where I get my hope and perseverance, among my other many adorable traits.

I wish that would tie my narrative up in a giant, neon-bright, happy endings bow. Despite all I've shared and without an ounce of sarcasm, I still enthusiastically believe we are the cutest, sparkliest country—like, ever! I still believe we have the best military in the world. And I believe more women should join the military. The more women we have serving, the better our already outstanding Armed Forces will become. Most of all, the more women join, the sooner we can reach a critical mass so meaningful, so undeniable, the paradigm of sexism, misogyny, sexual assault, and victim blaming will shift. I'm still an unswayable idealist and optimist in this regard. It will take a lot of patience, boldness, and perseverance, of course. But I believe this shift is inevitable—just as strongly as I believe in the power of words.

They still don't come easily though, them words. Even as a published writer, as a teacher and tutor of writing, a copywriter and editor, as a translator, an expert conversationalist, and all the other hats I wear for leisure or because a girl's got bills to pay—my goodness: writing's a %&*#@ process!

Clearly, it's a love-hate relationship. I still write though, mostly because silence feels like an enduring foe to human connection. I believe words, difficult as they are to come by, are indispensable tools in the worthy quest towards developing significant bonds. Everyday I feel it and I see it, how finding the words—finding that right word—is usually worth the struggle.

Essays

Essay Winner

Lauren Stevens

Georgia On My Mind

"Take her and hide!" Having plucked me from my warm bassinet in the nursery, the nurse rushed into my mother's room at the military hospital, panic evident in her face. She thrust me—black fuzzy hair, chubby fists, legs and all—into my mother's arms for safekeeping. Understanding the urgency and gravity of the nurse's request, my mother immediately launched into protective mode, shuffling, with her back hunched over me cradled in her arms, mustering all of her strength and channeling it to her recovering abdomen. Shocked, reacting through instinct, my mother locked herself into a bathroom stall, willing me not to utter a peep, lest we be discovered. After what seemed like hours, my mother emerged from the bathroom and was debriefed; a G.I. had entered the clinic with a gun, demanding his records and threatening all within earshot.

*

My life began on Robins Air Force Base in Warner Robins, Georgia, and the pull to revisit my roots and connect with my childhood homes has me driving with my retired father in an electric blue compact car, across two hours of countryside, from Hartsfield-Jackson Atlanta International Airport. Along for the journey, my father will serve as both my guide and my access to the Air Force base we'll be visiting.

As my husband and I begin to plant roots with our 5-year old son, and I near the end of my 30s, I feel compelled to pick up the collective pieces of my childhood, hold them in my hand and appreciate them through the lens of an adult eye. This particular trip is part of a plan to reconnect with each of my childhood homes as I begin settling down and settling in with my own little family. Having spent my childhood continually shifting across continents, and much of my adult life moving in tandem with career climbing, I'm finally reaching a point of stillness and reflection.

*

Roughly two hours south of Atlanta, beyond the jam-packed highways and sprawling suburbs, lies an old boomtown, deep in the heart of Georgia. Dodging cars on the highway, the road opens up about an hour or so south, with views of vast fields and wild country, before we begin seeing signs of life outside of Macon. Like DNA scattered around a crime scene, every move a military brat makes, a piece of them is left behind; I'm here to see what I can retrieve in Warner Robins.

Stopping at an IHOP along the way, I unfold my legs from the crammed compartment to stretch and take in my surroundings, the sun hidden behind wispy gray clouds. The bitter wind cuts through the thin cotton of my t-shirt, and I furiously rub my arms, realizing I've packed for the wrong weather. It's cold here in March, and I've packed nothing but t-shirts and jeans, expecting the Georgia weather to be akin to Florida, all sunshine and seventy degrees. That I have absolutely no idea how to dress for the weather in my birthplace is telling. This place that holds such significance for me, the very roots of my existence, is completely foreign to me.

*

Mary Edwards Wertsch, author of Military Brats: Legacies of Childhood Inside the Fortress, would likely call this need to revisit my childhood homes, before settling with my family, part of the psychological diaspora that is common in military dependents who have spent their childhoods uprooted. A military brat herself, Wertsch so eloquently states, "As adults most of us manage to slow or stop the moving—and yet we still find ourselves caught up in a strange migration. It is a migration of the soul, all the more mysterious to us because it has no clear origin and no certain goal."

An intense gut feeling is driving me to visit my childhood homes, a feeling I suspect is largely motivated by my age and the comfortable stage of my career. That visiting all of those foreign countries and American counties will somehow allow me to fully appreciate the colorful, cultural, and mobile military childhood that was my norm, and to share them with my own little family. If life is viewed as a circle, my circle will remain incomplete until I make my way back to my childhood haunts, if only just to make sure those places really exist.

This draw, this need for concrete evidence, is part of the affliction of a transient childhood, and common among military brats like myself. Having interviewed hundreds of military dependents, Wertsch conclusively states that, "Military brats also have a penchant for building symbols of stability into their lives. For them it is not enough to be anchored in a relationship or a career—there must also be concrete evidence of stability, something they can see and touch for reassurance."

*

And here I am, the gooseflesh on my arms standing at attention, ready to experience the place where it all began for me: Warner Robins.

A town that had resonated only as two words on a piece of paper, a security question for an online account, or an obligatory response to the question, "Where were you born?"

I dread being asked where I'm from.

*

Every stage of my life has been peppered with the seemingly innocuous question of my roots, but for myself and many other military dependents, the question can be complex—not everyone wants to hear our long and drawn-out explanations. And as Wertsch says, there's just no way to answer the question of where I'm from without associated awkwardness. It's awkward because the question, when asked by a civilian, immediately sets me apart, when what I really want is to experience a sense of belonging.

Says Wertsch, "A civilian asks the question to get a handle on the person; but no answer the military brat can possibly give will satisfy. The let's-get-acquainted convention has begun badly, underscoring the gap between our world and theirs."

*

I know nothing of my birthplace, just the words I utter when prompted, something I've come to find strange as I've grown older. In contrast, my husband regales my son and myself with stories whenever we visit family in Delaware. As we drive through the streets of his hometown, he points out the homes where he and his closest childhood friends lived, the lake where he lost his front tooth playing ice hockey one winter, and the restaurant his friend's father owned. Everywhere we drive, there's a story and a connection, one that brings smiles to my husband's face as places trigger memories of his childhood.

Until now, all I could offer was that I'd passed the highway exit sign for Warner Robins on the drive down to Florida from Pennsylvania, yet another relocation I'd initiated in my twenties, in an effort to find myself and my place in this world. I would remain rootless until my 30s.

*

Having spent my childhood in constant motion, I'm programmed to want to move every two to four years. I attended three different universities before earning my undergraduate degree, and then moved for better jobs, more or different opportunities, and sometimes, just because.

Mary Edwards Wertsch proposes that military brats experience a delayed maturity, as a result of their rootless childhoods. Indeed, indecision played a part in my many adulthood moves, as did a change in college majors, career goals, and even the careers themselves. Of this, Wertsch expounds:

"Arriving at a consistent self is very difficult for military brats; it is antithetical to our entire experience. Moving constantly is about changing self constantly. We grow up continually rewriting our personal scripts in order to fit in. We are social chameleons, and we know it. We also know that someday that has to stop, and that learning how to stop is going to be painful…."

*

In one of my adulthood moves, I mistakenly thought that buying my first house in Florida would force me to plant roots. Too far off the highway to visit, I can still remember the white letters against the fading green of the highway sign as I sped through the state of Georgia, craning my neck for a longer look as I sped past: Warner Robins. I'd be heading right back up I-95 a year-and-a-half later to attend graduate school in Erie, PA—even home ownership wasn't enough to keep me planted in one place.

It's a curious thing, longing for a place you don't remember.

*

Unfolding myself from the car, I slowly walk toward the small brick building. Thankfully, it's sunny today, but March winds prevail and the only sound is the skittering of fallen leaves across the paved roundabout set in front of the hospital. To call this a hospital is generous, at best.

Utilitarian, nondescript brown letters, almost blending into the brick wall to which they're affixed, announce that this is the 78TH MEDICAL GROUP. This can't be right, the place before me is a single floor clinic. I try to imagine my mom undergoing surgery for a Caesarean here, let alone space for a nursery and rooms for patients recovering from procedures. I can't fathom enduring labor in this building, with thin walls and such an intimate space—there's no room for me to imagine any privacy. My mom's frightening story about the gunman in the hospital becomes that much scarier now that I've seen how small this building is.

A walkway leads from the front doors, across the roundabout drive, and into a circular island of lawn, flat and absent of any ornamental plantings, as bare and barren as the office parking lots on base this Saturday morning. The cement walkway opens into a paved gray octagon, where a pair of wrought iron benches face each other; I envision my dad stepping out here to smoke a cigarette after being told that he had welcomed a daughter.

*

A month before my visit, it's announced that this particular building, the oldest part of the medical clinic on base, is to undergo a twenty-nine million dollar renovation. Built in 1967, just ten years before I was born, the 78th Medical Group building's present-day use is for administrative purposes, with hospital rooms serving as offices, each with its own bathroom.

This kind of repurposing is common on military installations. I grew up attending services in a makeshift church on an old RAF base in England, first in an old storage building, later moving to a larger space in the ballroom of a WWII hospital. My K-8 school in England was composed of a series of office buildings. Military families are a resourceful lot; we make do with what we've got and what we're given.

Built in the 1990s, the newer 78th Medical Group building is more of what I'd imagined, with an automatic entryway door, multiple floors and a decidedly warmer exterior. The timing of my visit couldn't have been better; I've gotten to see the building before it becomes unrecognizable and nothing like it was in 1977.

Shortly after my visit, this building will be gutted, removing everything down to the support beams, topped with a new roof and brick to match the newer hospital looming large in the background.

Every change on military installations mars the landscape of memories for thousands of military dependents like myself. Each closure, each demolition, means that a home, a part of our past is no longer there, no longer able to be visited or experienced as an adult, no longer there to complete the circle of connection.

*

I'm certainly not alone in wanting to retrace my roots and reconnect with the many places that make up my childhood. On any given day you'll find numerous people posting in military brat Facebook groups, rounding up people stationed at the same military bases, shout-outs to favorite stations, reminiscing and revisiting, now that they're adults. I belong to two military base-specific groups, where we share in the excitement of those counting down a visit to Hahn, Germany or RAF Little Rissington, England. Those with shared stations covet the photos fellow brats post, some even ask for their former houses to be photographed, anxious to see if those glimpses match the memories, forever etched in their minds.

However, the end of the Cold War, and the fall of the Iron Curtain, meant that many military installations across Europe, and even in the United States, were closed due to budget concerns or deemed nonessential. In fact three of the five military bases on which I spent my childhood years are no longer open; one has been repurposed into a local village in England, another sits abandoned, slowly deteriorating behind chain link fencing in the German countryside, while the other base, Stateside, had a brief moment of revitalization, when 1999's iteration of the Woodstock music festival was held on its flight line.

*

The fact that the 78th Medical Group's building renovation will displace office workers for at least two years, forcing them to work out of a temporary modular building, isn't lost on me. To be a member of the military, or an attached dependent like myself, is to embrace change and impermanence as a way of life.

After visiting the hospital, we head to the other side of the base to view my father's old domain—the flight line. As an air traffic controller, my dad spent entire shifts high in the tower, alternating his view between the vast, gray concrete of the runway and the complex, lighted dashboard of the instrument panel. As an Air Force brat, I've spent a lot of time on flight lines.

We park at the edge in a far corner of the tarmac, which is lined with tall loblolly pines. We stay on the edges, so as not to alarm security personnel, taking in the retired planes parked upon the grass, and looking out across the straight lines of the vast runway. My father takes a moment to give me a short lesson on the ins and outs of runways, using the largest runway in Georgia as his guide. It's like riding a bike for him, and I'm sure there's a part of my father that feels at home when he's at an airfield.

When we were stationed at Robins Air Force Base in the 1970s, Jimmy Carter was president and Air Force One would be housed here each time he came to visit his home in Plains, Georgia. My father demurs when I ask if it was a big deal when the president landed, he says it just meant there were a lot more people around.

I take a couple of photos of my father out on the flight line, with the control tower in the background. We've seen what we can see here, and my dad has told me all he has to tell. It was just work, a job, in his mind. For me, it's a piece of my own personal history, something I couldn't have possibly understood or appreciated as a child, but hold dear now that I'm an adult. While moving from base to base is a normal part of a military childhood, it's interesting to explore why my father was sent where he was, now that I understand his career in the context of political history.

As we drive away from the flight line, I stop the car by the line of pines. My father shoots me a quizzical look, I sheepishly tell him that I want to grab one of the oversized pinecones that litter the ground beneath the trees. Larger than my hand and stouter than most pinecones I see up north, I snatch two perfect specimens, the end of each of the bud scales tipped with a thorn-like protrusion. I handle them carefully and hurry back over to the car. Hopping in and putting the car into drive, I have a sense of fulfillment, knowing that I have something tangible from inside the base gates, from the very soil of my place of birth.

As we head back to the hotel, my eyes flit between the rearview mirror and the road as we pass through the gates, the entrance getting smaller in the distance. I've seen all that I can see here.

*

Travel writer Pico Iyer gave a wonderful TED talk in 2013, entitled "Where is Home?" In his talk, Iyer discusses his own mobile childhood and wrestlings with both cultural and national identity, but he eloquently summarizes the plight of our increasingly mobile society, and more specifically those with rootless childhoods. Iyer says, "Their whole life will be spent taking pieces of many different places and putting them together into a stained glass whole. Home for them is really a work in progress."

Turning the pinecone in my hand, I get lost in the seemingly endless swirls of woody bud scales, fanning out from its central axis. Nothing is arbitrary in a pinecone's design. One of the many examples of the Golden Ratio in nature, my pinecone illustrates the Fibonacci sequence, with each number the sum of the two preceding it. My birthplace doesn't define me; I am the sum of the many places I have lived.

Shane Griffin

Accidental American

I've heard people say to leave history where it belongs—in the past. And I've heard people say we have to study history so we never repeat its mistakes. Since I was a little boy, I have been fascinated with history—especially military history. My prize possession was a full set of *Encyclopedia Britannica*, a gift from my grandmother to my mother. My mother never touched them, but I read them every day. My puny arms struggled to pull the volumes out and set them on the floor. Sometimes I knew what I wanted to look up, and other times I just browsed through the table of contents and let my curiosity take me anywhere in the world or anytime in history, an early form of web-surfing.

My grandfather on my mother's side was born in Germany. I was fascinated by this. I didn't understand why he didn't fight for Germany in World War II. I was too young to understand world politics and immigration, but I tried anyway. I consulted my encyclopedias. I was fascinated with war. And about that time, when I was trying to figure out the world, *Star Wars* came out. It was the first movie I saw in the theaters. I was glad Luke, Han, and the rebels destroyed the Death Star and the Imperial Army. But there was something that fascinated me about the Imperial Army and the Dark Side. I think it was the uniforms, or just being so powerful that the entire universe was under their control. At night, I would walk around my yard and look up into the sky. At that age, it was hard for me to understand the reality of things, and a piece of me wanted to believe that up in the night sky, in a galaxy far, far away, there was a struggle taking place between good and evil. But all I had to do was look into what was happening here on Earth and in my own family to find those kinds of struggles, divisions of ideology, and complicated political boundaries.

It took me a long time to come to realize that Germany had lost WWI and WWII. I don't know why this bothered me so much. I liked their uniforms. They looked like the *Star Wars* Stormtroopers. I read about concentration camps, but didn't understand that either. I was raised in confusing political and social boundaries. I got into a fight at school once because I talked about my grandfather coming from Germany, and a boy told me that all Germans were pussies. I got sent home from school after I punched him in the face.

I still grew up trying to figure out how my grandfather was born in Germany yet still fought for America in WWII. You wouldn't think my grandfather had any German in him at all if you saw him on the street. After the war, he returned home to Coon Rapids, Iowa, and farmed with his father. Then he left farming and ran heavy equipment until he retired. People called him the Catman, because he always wore a black baseball ball cap with a CAT patch on the front.

During family gatherings, the German would come out of him. After a few drinks, he broke out his accordion. He played it and danced around the living room to entertain all of us kids. While he played the accordion, he kept his eyes closed, and he danced without stepping on anyone or knocking anything over. Then he sat in his recliner and played his harmonica, eyes closed, as tears rolled down his cheeks. He liked to play slow songs. His hand would quickly fan the chords, and the notes bounced around the narrow living room walls and across mine and my cousins' little eardrums. The rest of the adults were in the kitchen playing cards. My grandfather stopped playing. He set his harmonica in his lap and pulled a handkerchief from his pocket and wiped the tears from his eyes. He was bound to another place by music and song. I always wanted to know why.

His parents, my great-grandparents, lived in Denison, Iowa, after they retired from farming and sold their farm. I would have no choice in going with my grandfather to visit his parents; my mother made me go. The trip from Nevada, Iowa, to Denison seemed like it took forever. They lived in a small house on a hill. My great-grandfather, Henry, always sat in the front room in his recliner reclined all the way back, facing the TV. Professional wrestling was always on. My grandfather would walk past his father. "Hello, Dad." No response. "Hello, Grandpa." No response. I asked him about why Great-Grandpa never talked to me. He said that where he came from, grandkids were supposed to be seen and not heard. Thankfully, my grandfather didn't follow that rule. He took me fishing, and we went on trips together, and he would play with me and all of my cousins. I remember him holding me as a young boy. His rough face against my soft skin as I snuggled close to him, the smell of his aftershave. Electric Shave—more alcohol than lotion.

My great-grandmother was another story. She would talk to me and liked being around children. She spoke English well, but with a very strong accent. It was so strong, she rolled her R's like Spanish speakers. I learned later that their accent was called Low German, the accent of the commoners. Hitler and most of Germany during WWII spoke Low German. If you watch speeches of Hitler on the Internet, you'll notice he rolls his R's a lot. It's ironic now how modern Germany speaks High German. During the war, High German speakers were used for radio transmission among the Wehrmacht, like we used Native American code talkers. And most younger Germans today need a translator to understand Hitler's speeches. It is strange how inside of one small country, small things such as accent can cause confusion.

She would hobble out of the kitchen to greet us. "Halo. Come to kitchen, I have cheese und crrrrackers." My great-grandmother Louise was kind. A sweet little lady. Her kitchen was her domain, and she would play German records while Great-Grandpa snoozed to tacky 80s professional wrestling on the TV. I couldn't understand what the musicians were singing, but my grandfather did, and he ate his food and tapped his foot on the linoleum to the beat.

I was born in the time when color TV and color photographs started to become normal and not too expensive. But as I walked through my great-grandparents' house and looked at family photos, I wondered who all the people were. While my great-grandfather watched wrestling, and my great-grandmother talked to her son, I looked closely at all the photos. There was one of a young man by an artillery piece. He had a curled mustache and a spiked helmet. His cheekbones were high and his eyes squinty like my great-grandfather. Another picture had a black banner draped over it. The man in the photo looked like the Wehrmacht soldiers I read about. And there were many more photos of military men. Some were just headshots, but they were black and white, not showing arms or the chest where all the medals would be. I asked my grandmother.

She said, "This your great-grandfather. This one is my brother, and all the rest are brothers and nephews of your great-grandfather. Most of them have died. But this one is my brother. He never returned, and this is all that I have of him."

*

I tried to ask my grandfather over the course of his life about his past. I knew about the time since I was born, but what about before that? I knew that he emmigrated with his family from Germany in 1929 when he was seven years old. These areas in the Midwest were popular among German immigrants. My grandfather said his classes were taught in English, so he had to learn English fast. His teacher did not have the patience to teach foreigners, and she often smacked the children with the wooden chalkboard pointer she carried around. He and his sisters would come home from school and help teach their parents English.

I asked him if he remembered his grandparents. He said no. I asked if he remembered the day he left Germany. He said no. Maybe too much time had passed. He remembered when he found out about Pearl Harbor. He and his father were in Anita, Iowa, for a cattle sale. He remembered going to the local café and eating breakfast. They then went to the sale, and the auctioneer announced before the sale started that the Japanese had attacked Pearl Harbor and that there were many lives lost and ships sunk. He said he and his father didn't buy any cattle that day and they didn't speak on the drive home.

My grandfather wanted to enlist, but he was torn. He knew he had family members in the Wehrmacht and he might have to go and fight them. His mother had told him only sad stories from when Germany was in WWI. His father never talked much about his WWI service. In my grandfather's German family, every male that was old enough had to serve in the Wehrmacht, and the young boys and girls were forced to be in the Hitlerjugend. Every school in Germany had became a public, military-style school. All the children had uniforms to wear, were required to learn drills, camping, military skills. No one in Germany questioned the Nazis, because they had lifted Germany out of the economic depression, and to question the regime meant losing your job, having the State confiscate your assets, being kidnapped, or executed.

My grandfather "graduated" from school in eighth grade. That was normal back then. Only more well-off kids went on to high school and college. My grandfather was not a citizen when he was drafted in 1942. He was a child when he immigrated here with his family—an accidental American. If that happened today, he would have been a DACA child.

*

He reported to Camp Dodge, Iowa, for basic training, then on to Camp Cooke, California, for more training. On that train ride from Iowa to Los Angeles, he met Violet Finch from Kelley, Iowa. They talked for a while and exchanged directions on how to get a hold of each other. Violet was on her way to Los Angeles to work in the North American Aircraft plant. Ernest and Violet would meet up several times, before Ernest and his unit were then sent to Fort Dix, New Jersey, to be shipped out to England. He and his unit waited there until they landed in France on June 20, 1944—D-Day plus fourteen.

My grandfather recalled being stationed in England. White and black troops were segregated into different areas. His camp was on the other side of a town from the black camp. He said there was a lot of tension between the white and black troops when they were on leave. The troops were not allowed to be too far from their units. There was a riot one time, and the American and British military police had to put it down. The mayor of the town had banned all American servicemen from entering his town from that point on.

My grandfather landed in France, and he drove a supply truck all over France and Belgium. He remembered being out on the road when the Battle of the Bulge started. The attack first started with heavy artillery bombardment and German infantry. My grandfather narrowly escaped those first few days of the battle when the American lines were broken. But there was always the fear of the Germans breaking through the lines and cutting everyone off. Eventually the lines stabilized, and the Germans were beaten back to where they started.

His fears of hurting or killing a relative remained, and he had made it all the way to Aachen, Germany, which is across the border from Northern France, without incident. Still, this was nowhere near his hometown, Tornesch, which was about two hundred and fifty miles to the north. I asked him once how he would feel if he had killed one of his relatives. He told me they were the enemy, and he wouldn't think much of it if he had. But in an odd twist of fate, as soon as his unit touched German soil, they were given orders to report back to Antwerp and board a ship for the Philippines. While there, his unit processed Allied prisoners of war. His unit then was tasked with transporting supplies across Japan after the war's immediate end.

When he returned home, my grandfather married Violet Finch, the young woman he met on the train. They settled in Coon Rapids near his parents' place and raised a family. My mother was born on that farm.

*

I served in the military, including tours in Iraq. I had always wondered about my German family members who served in the Wehrmacht, and I had from time to time asked my grandfather what he remembered. But at that age he couldn't remember much. All that I had from him was a town, Tornesch. When I returned from Iraq in 2005, I had made it a point to not put things off anymore, to accomplish things I'd always wanted to, and not be one of those people who lived the rest of his life saying that he wished he would have done this or that. After nearly being killed several times, your perspective on life changes and you realize that you have only so much time in this life to have fun, but also to do the important things. So I set off to find my German relatives.

Using Google, I started to track down family members. My grandfather couldn't remember specific names, and I had lists of Ludemanns. Finally, I found a German phone book for Tornesch, and I started at the top of the list and began to call numbers. Some people who answered talked to me for a while until they realized we were not related. Others hung up on me, and then about halfway down the list, I got in contact of one of my great-aunts who spoke English very well and knew exactly who I was.

We corresponded several times, and I convinced my grandfather to go along with me, but in the end he stayed home. When I asked him why, he said he was afraid of what they might think of him since he served against them. I assured him that that was old news and they would be glad to reunite. But those old feelings of fear and guilt are hard to forget, so I went without him.

My great-uncle, Joern, met me at the airport in Hamburg. He spoke clear English. He had been a Merchant Marine and traveled all over the world. He was Danish by birth and married my great-aunt, Birget. He was a well-educated man, fluent in five languages: German, Danish, English, Spanish, and French. He was also the self-titled family historian. He had binders full of information. He had spent countless hours researching and building the family tree, which had grown to twenty pages, and it took fifteen minutes to line up all the pieces of paper. It literally covered his entire living room floor, with family members dating back to the 1600s.

I was only there for a week. The history was so abundant and rich and my relatives were so full of stories that I didn't have time to collect all of them. I enjoyed those stories about the family, but my real interest was in my family's military history. I asked my older relatives about it because I had ignorantly thought that the wounds would have healed.

My uncle Meinhard went into his bedroom and brought a big box and set it on the kitchen table. With Joern present to translate, we went through the pictures. There were many black-and-white photos of my family members, the men in uniform. One thing I noticed right away was all of the pictures were copies. The photos were cropped in such a way as to remove the Nazi armband with its notorious swastika, and they reminded me of my great-grandmother's pictures that didn't include the soldiers' arms. I had to steady myself for what was coming. It was obvious they had served in the Wehrmacht, but I had hoped that none of my family members were death camp guards. Thankfully, none were. Like me, they were fighting men. And I knew that whatever ideology brought you to wear a uniform, when you are on the front lines, it is really about you and your comrades in that moment in that area fighting for your lives.

Nearly every male in my German family fought for the Wehrmacht. They didn't actually know what unit they belonged to, but the stories remained. The oldest of my relatives to fight for the Wehrmacht was my grandfather's uncle, Wilhelm. He was involved in the Siege of Leningrad. He caught frostbite so bad that the military sent him home. He refused to have his leg amputated, and dealt with the pain for his entire life. Meinhard recalled watching his mother remove, clean, and redress his wounds every night. He said that you could look into his body and see the muscles and tendons moving. He was sick a lot and had to visit the hospital frequently, but he was always treated with a homemade elixir that was supposedly a cure-all. It looked like cloudy whiskey and smelled like Jägermeister. My aunt told me that I could drink it for stomach ailments, or use it as a sleep aid. She moistened a cotton ball and gave it to me before I went to bed that night, said to put it in my ear that rang constantly with tinnitus, an injury I sustained during an IED attack in Ramadi. I woke the next day with a clear mind, and although the tinnitus was still there, it had lessened in severity.

I learned that one relative was a pilot in the Luftwaffe. He was killed in the Battle of Kiev when his fighter plane was damaged and he crashed into a factory's smokestack. Another distant cousin was in the Battle of Stalingrad and was captured. He survived only to spend the next five years in a Soviet Gulag. Once he returned to Germany, he was a different man and drank himself to death a few years later.

Klaus, my grandfather's first cousin, had fought in France and was in Germany fighting the Americans when the surrender was signed. He was disarmed and sent home on foot. The family story has it that they all knew my grandfather was in the American army and possibly in Germany, and he stopped at every American camp on his way home and asked for my grandfather by his German name, Ernst Ludemann, but he was told to stay out and was often beaten back by unfriendly American MPs.

Another of his first cousins, Kurt, had also fought on the Western Front and survived the war. He immigrated to Canada and worked in the forest, cutting down trees for the *Chicago Tribune*. He had to work on the logging crews for five years before he earned his Canadian citizenship. He then married a Canadian woman and lived the rest of his life in Windsor, Ontario, as a factory worker. I had corresponded with him several times by phone and letter and even arranged a visit with him, only to have him cancel because of health reasons. He knew that I was interested in his story, but I wonder if he was too ashamed to share it. He told me that one time my great-grandfather had come back to Germany to visit, and he had convinced him to immigrate to America. They both went to Hamburg to the American Embassy only to find out that America was not taking any more immigrants that year. But the Canadian Embassy was next door, and they both went there and found out about the logging crew opportunity. He left Germany a month later.

Others, who were too young to fight for the Wehrmacht, were in the Hitler Youth. My uncle, Meinhard, was one of them. He said he had to wear a uniform to school every day, and close-order drills and physical training became a part of their school curriculum.

*

When I returned home, I shared the photos and souvenirs I brought from Germany. My grandfather and I sat for hours and talked about his birthplace. I could tell as he sat across my kitchen table from me that his mind trailed off from time to time, the synapses of memory connecting intermittently. It would have made better sense to him, but I was grateful that I actually got to step inside his childhood home. I had seen pictures of it before, and it hadn't changed much since the pictures. It was a house connected to a barn. The barn still had a thatched roof. The current owner was gracious enough to allow me to enter the house and take some photos. I showed them to my grandfather, but none of the interior photos jogged his memory. The outside of the house and barn did, though. His eyes trailed off somewhere behind me, out the big picture window in my kitchen.

"You know, Shane, I should have just gone with you."

"I know. But it's too late now." I didn't know what else to say, and I felt bad about saying what I did.

*

I have never returned to Germany, although I wish I had. The requirements of family life take priority over numerous overseas trips. Over the years since my trip, my grandfather would call me out of the blue and say, "Shane, let's go to Germany. You and me. I'll pay for it." I was excited to hear him make the offer because I thought it would be amazing to see this place with him, but those promises never materialized.

Some years later I skyped with our German family and arranged a time for them to talk with my grandfather. I set the laptop up on my kitchen table with my grandfather at my side as I dialed them. They answered, and I explained to Grandpa the different screens and told him not to shout. He thought they couldn't hear him. Joern began to talk in English and soon started to talk in German. I watched my grandfather struggle with some of the words, and then he began to speak German fluently and carry on a conversation with Joern. I was impressed that my grandfather could recall his native tongue, an immigrant who could have had a different life in a different country. I was reminded of my youth and visiting my great-grandparents. A part of me felt cheated that I didn't grow up knowing more about my heritage. But I understood the confusion and the desire to assimilate into a new culture and identity.

I sometimes wonder what it would have been like if they had not immigrated when they did. I wondered if my great-grandfather knew what was coming and got his family out of Germany before it was too late. If they stayed, I wondered if my grandfather would have served in the Wehrmacht and been killed or injured. Or even if I would have existed in the first place. It was dizzying to think about. But I still imagined myself and the rest of my family in this alternate present, affected by events in the past that never happened.

I am a legacy, the second generation, of an accidental American. A young child of immigrant parents. I had stood on that dock in Hamburg, probably the same dock my grandfather walked on to the SS *St. Louis*, which took him to Ellis Island. I wondered if he hugged the rail and waved at his family for the last time. I hope he did, because that little boy never returned to his homeland.

Ten years later, I said goodbye to my grandfather. I wasn't able to unlock any more memories before he passed. But while my family and I were cleaning out his house, we found letters he had written to his family and to my grandmother while he was overseas. The letters are filled with excitement, homesickness, musings about life, and the fear of dying at any moment in combat, and of course—love. They read like any American kid would write home to his parents and his sweetheart. And maybe that's all that I really needed to know about my grandfather's accidental American citizenship: he became an American, and lived just like any of the rest of us, finding his own version of the American dream.

He spent nearly his whole life here. He became a citizen and did not take it for granted. He served his country, and left his family behind to wonder and worry. He will live on in my memory, and I hope in the rest of my family's thoughts and prayers. My German grandfather loved and cared for me and his family. He was gentle to his enemies, and in the end, he slipped away in peace while reaching for something in the air that I could not see, something just outside of my plane of sight and understanding. I'd like to think it was God. Or could it have been his grandfather reaching out for him, too? I hoped so.

*

That year, 2017, my own grandson was born in July. My daughter was gracious enough to allow me into the room with her while he was born. My wife and daughters stayed in the hospital room to help take care of the baby, making sure they didn't miss any new moments of his young life. I went home that night by myself.

I woke the next day before the sunrise. I was anxious to get to the hospital and see my grandson in his first full day on this earth, in this new life of his. But I had paused to let God show me the hot orange sun break the horizon and rise into a perfect circle low in the morning sky, and I wondered if my great-grandfather stopped to see the sunrise on my grandfather's first day on this earth. And as a grandfather, I want to be just like him, because that is the only way I know how to love my grandson, like my grandfather loved me. I took a picture of Weston's first sunrise on my phone, and one day I will show it to him. He may not know what it is or what significance it carries for me, but maybe someday, after he has lived the dramas in his life and had his own family, including his own grandchild, maybe he will think of me and the distant promise of a bright warm sunrise.

Bill Glose

What it Means to Serve

For most of its history, the bucolic hamlet of Bedford was no different than any other rural American town. Founded in 1839, Bedford blossomed in the shadows of the Peaks of Otter into a production and distribution center for tobacco, corn, wool, and other locally grown products. The town grew slowly, adding a mill here, a storefront there, but nothing much really changed. Farmers sowed and harvested with the changing seasons as had countless generations before. The people here were the strong, silent types who spoke their minds through action. When a fire burned a neighbor's farm, the community didn't send sympathy notes; they banded together and rebuilt the barn. And when Pearl Harbor was bombed in 1941, every young and able-bodied man showed his patriotism by enlisting in the military.

Thirty-five of the "Bedford Boys" were assigned to the Army. This was the era of "cohort units," meaning that troops who signed up from the same town were kept together in the same unit. The intent was to provide a built-in cohesiveness and camaraderie. Thirty were assigned to Alpha Company, 1st Battalion, 116th Regiment of the 29th Infantry Division, and the other five were assigned to other companies in the regiment.

What the cohort policy didn't account for was the devastating blow that could hit a single hometown if a unit took heavy losses. And no heavier losses occurred than on the D-Day invasion of Normandy.

On June 6, 1944, an Allied Expeditionary Force consisting of twelve member nations launched Operation Overlord, an amphibious assault across the English Channel against German-occupied France. The invading force included more than 11,000 aircraft and 5,000 ships carrying 150,000 men. In the first wave to land at Omaha Beach were the thirty Bedford Boys of Company A, whose unit was the "tip of the spear." They were the first to storm the heavily fortified coastline. And they were the first to fall.

Alpha Company took 96% casualties on Omaha Beach. Nineteen of the Bedford Boys lay dead and others were wounded. Those who survived pushed on through withering machine gun fire, bombardments, and wire obstacles until they established a foothold on the beach so that heavy equipment could land safely. As fighting continued in the Normandy region, four more Bedford Boys died, two from Alpha Company and two from other units. By invasion's end, only twelve of the original thirty-five remained alive.

For a town as small as Bedford to lose so many of its sons was a tragedy beyond comprehension. Their community's loss was the severest per-capita death toll of any battle since the Civil War. To honor the memories of those who sacrificed everything, local citizens petitioned the federal government in 1988 to build a D-Day memorial in Bedford. When Uncle Sam refused to back the project, residents formed a non-profit called the National D-Day Memorial Foundation and spearheaded a fundraising campaign. One of the largest donors, with a one million dollar gift, was Charlie Brown's creator Charles Schultz, a WWII veteran who saw combat at the tail end of the war. (In typical Schulz fashion, he once commented that the lone opportunity he had to fire his machine gun was missed because he had forgotten to load it.) By 1994, the foundation had raised $25 million and construction of a multifaceted monument began on an eighty-eight acre parcel set high on a hill overlooking the town of Bedford.

My personal connection to this town and this memorial dates back to the 1980s. To help pay my way at Virginia Tech, I joined the National Guard and was assigned to the Bedford unit, Company A of the 116th. Once a month I participated in weekend drills that began in the armory (a converted gymnasium) and often took us out into the Jefferson National Forest. When we formed up at the beginning and end of the day, I would marvel at our company's standard, which bore a cluster of colorful battle streamers, including one embroidered with NORMANDY.

A high school buddy of mine, Brad Lawing, followed a similar path. A year behind me, he also came to VT, where we roomed together for a year. He joined the National Guard and was assigned to a company in Roanoke. And after graduation, he was also commissioned as an officer in the Army, although he made a career out of it while I got out after five years of active duty. In 2010 I attended his promotion ceremony and watched Brad's commanding general pin the silver oak leaf clusters of a lieutenant colonel onto his collar. Brad's once-brown flattop contained more salt than pepper, but otherwise he resembled the same athletic, bullet-headed poster boy for the Army that he'd always been.

Later that same year, Brad and I hiked nine miles from Thaxton into Bedford to visit the National D-Day Memorial. I wore a rucksack while Brad slung a two-quart Camelback over his shoulders. The forecasted rain held off, and fat, black clouds hovered overhead, providing us with shade. We walked at a brisk but comfortable pace and sang a few jodies to match our cadence, and chatted about our respective time in the Army. The entirety of my short career had been in the infantry, both as an enlisted soldier in the National Guard and as an officer in the 82nd Airborne Division. But Brad had been commissioned as an armor officer, commanding soldiers that operated tanks.

"I never wanted to be a tanker," Brad admitted. "When I got branched armor at Tech I kind of tried to fight it. But then I got to my Armor Basic Course and I started going through it and tanks seemed kind of cool. The last thing was the ten-day war [game] where you're out doing missions and you live on your tanks. And from that point on, I thought, 'Tanks rock, man!' You muldoons pull out a little piece of a map with an eight-kilometer movement, and you're whining because that's an all-day, all-night thing and you're going to be dead tired when you get done. But for me, eight klicks is like five minutes. And when we get there, there's always a shivering infantryman who comes up beside our tank and says, 'Um, can you turn your tank on? We're cold and we want to stand behind it.' We do a lot more maintenance, but we fight in style."

Although Brad's time as an officer had been in the cavalry, his time as an enlisted man had been in the infantry, in both the National Guard and during a two-year hitch with the 3rd Infantry Regiment in Arlington, VA. Their nickname, the Old Guard, stems from the fact they are the ones who guard the Tomb of the Unknown Soldier, a duty Brad performed over an eighteen-month period that he still reflects on as the best assignment he ever had.

"What was so special about it?" I asked.

"Because I was able to give back to four guys who gave everything they had—name, identity, *everything*—just to protect his brothers in arms, and all of us back here and our way of life.

"I remember being out on Memorial Day morning at six o'clock. Because of the sunrise service we were out early in blues that morning, so I went out for the first walk before the cemetery opened up in blues. It was a standard walk, which we didn't normally do when the cemetery's open, but we do it now. So it's just me, the sunrise, and my four brothers, and nobody else; and that's the highlight reel right there. It's not about the crowds. I mean, it's kind of cool when people take your picture, but it's so much better when it's just you and your brothers there. That's a fond memory."

"So what was a typical day like when you were guarding the Tomb?"

"When I was there, there were three reliefs based on height because you wanted the guards doing the guard change to be about the same height; you didn't want a really tall guy switching out with a really short guy. So six-foot to six-foot-two was third relief; second relief was six-foot-two to six-foot-four; and then third relief was everybody over six-foot-four. So I was on second relief. And at the time we were one twenty-four-hour shift on and then two days off. And the first off-day was a practice; you would go in and you would practice all the stuff, work on the uniform, and then you go get a haircut. And then the second off-day was completely off. And that was the rotation. But when you were on, you would be down at the Tomb for a twenty-four-hour period. The cemetery is open only eight to ten hours, so when the cemetery was closed we had slightly different procedures for the guy outside guarding. But there was still a guy there twenty-four hours a day, seven days a week.

"The guard changes every half-hour on the hour during the summer and every hour during the winter. During the summer, you might do three or four back-to-back half-hour shifts then take a two-hour break where you hydrate and fix your uniform from all the sweat spots. On a bad day—let's say you got a guy on sick-call or injured or leave—and there's three guys down there, one dude is doing the changing and two dudes are going back to back for twelve hours. Let me tell you, I've done that once and I think I lost ten to eleven pounds that day just sweating. Because you've got to figure, a guard change takes seven or eight minutes. So after the guard change, you go downstairs, take the uniform off, drink about a half-gallon of water, try to re-polish your shoes—they'll fade in the sun and they'll crack, so you've got to get a little bit of shine back. If you sweated on any of your brass, you've got to wipe that off or the sweat will pock it; it'll just eat it up, tarnish it, especially during the summer. You sweat at your elbows, your knees. And then at a quarter-till, you start getting dressed again to go back outside. So that gives you about a seven- or eight-minute break until you have to go back outside. Not a lot of time."

"It must be difficult to keep your composure while you're on guard."

"Oh, yeah. This one day I was out there during the summer, and people were milling around before a guard change. A little boy walked up to the chains, and the chains at the time—they're different now because of 9-11; they've pushed everything back—but at that time, people could get within five or six feet of the guard. They could almost reach out and touch him. And I'm standing there and this little boy came up—he's probably six years old—and he starts talking to me. And I can't say anything. And he's like, 'Hey, mister, where you from? Hey, mister?' And he's asking me a thousand and one questions and I'm trying not to laugh. I mean, it's all I can do to keep my count going and not laugh. So, when I start going down the mat, he starts walking beside me. He's mimicking me and I get down to the end and he's asking his questions, and this went on for fifteen minutes. The entire time, he was like, 'Aw, come on mister, say something to me.' A couple of people were giggling. And it was harmless. I had no issue with that. And then finally someone grabbed him when the Sergeant of the Guard was coming out for the guard change.

"And then another time there was this World War II vet being pushed in a wheelchair. Both he and the guy pushing had their VFW hats on, and they came up to the chains and they just stood there for about ten minutes, didn't say anything. And the old dude in the wheelchair tried to stand up—he couldn't quite do it—but both of them rendered a perfect salute, dropped their salutes, and then before they walked off, they got as close to me as they could and said, 'Thank you for what you do.' And *that* right there reminds you what it's all about. I mean, it ain't about me; it's about the guys who gave it all, the four unknowns. It's hard to explain to civilians why we believe what we believe and why we do what we do, but there's a brotherhood in the military that just sticks with you. You know what I mean?"

"Have your children been there and seen what you used to do?"

"Not yet. I'm looking forward to the day when my boys are old enough to kind of understand it so I can take them up there and show it to them."

"Your wife would probably be glad if you were still in the Old Guard."

"Yeah," Brad said, "Rebekah would probably go for that."

Brad's wife was a Southern belle from South Carolina with perfect teeth and long brown hair. At that time, they had been married for eleven years, during which Brad had left Rebekah stranded six times while serving on hardship tours—one in Haiti, one in Bosnia, and four wartime tours in the Middle East, the last of which Brad described as "fifteen wonderful months in Baghdad." And he'd just received news that he would be leaving again for his second tour in Afghanistan.

During my father's twenty-three years of service in the Air Force, he had served on two hardship tours, one in Vietnam during the war and the other in Iran just before Ayatollah Khomeini rose to power. I was too young to remember life on the homefront while my dad was at war, but I recall how his year in Iran affected our family. My mother barely slept and began to smoke Virginia Slims. We all huddled around the TV to listen to reports of rioting students and fears of an impending coup. That was the single hardship tour that I recall, and it was damn hard on everyone. I couldn't imagine anyone being sent on *seven hardship tours* in such a short time. But such was the world we lived in, one with shrinking manpower and two seemingly endless wars.

"Rebekah handles it well," Brad said. "But I hope this is the last one. She's put up with a lot."

"Does she come from a military family like we did?"

"Nah. She had no experience with the military and knew nothing about it. Three months after we were married, after I sent her to Fort Stewart, I was gone to NTC for a month. And then I deployed to Kuwait. I was there for our first anniversary. So, yeah, she's put up with a lot."

"How long is this tour?"

"Six months. Supposed to get us back in time for Christmas. But you never know for sure until you actually get home."

"All these times you've been over there, what's your opinion of the Middle East?"

He blew out a puff of air and said, "It's hot."

I laughed. "Anything else?"

"It's hot *and* it's sandy. It's like sitting in front of a dryer exhaust vent all the time. And the sand is not like beach sand; it's like talcum power. It's all this really, really fine nastiness that gets into everything. No matter what, you're always covered with dust. No matter what you try to do, you get dust in your eyes, and in your ears, and in your mouth, and in every other crack you've got, and it just turns into a nasty paste and you just live with it. So, yeah, it's just hot and sandy."

"What's been your most vivid memory from your time over there?"

"Some of the guys we met. There was this one dude, he was a master sergeant in the Afghan Army. He couldn't have been five feet tall. He had this big beard and we called him Yard Gnome—'cause he looked like that yard gnome in the Expedia ads. But, man, you should have seen him in action. He would be telling these privates what to do and they would be all lackadaisical, and he would just get a stick and beat the crap out of them. He would be yelling and screaming and just whacking on them."

He then told me about an argument he had with an Afghani tanker who had been fighting on tanks since the age of fourteen. The Afghani claimed the maximum effective range of his tank was nearly three times that of the U.S. Army's cutting-edge M1A1 Abrams, and Brad told him that was impossible. But then an interpreter got involved and they realized they were talking about two different methods of firing.

"They wouldn't use the sights because they don't use tanks like we do in a direct-fire mode," Brad said. "They support the infantry with tanks in an indirect-fire mode. So they'd be sitting in the gunner's hatch with their head out of the hatch and then they'd fire a round [like a mortar] and see where it hit and then manually adjust and fire another round. And [this Afghani] would be pretty much on just because he'd been doing it for so long.

"Let me tell you something, he knew his stuff. He fought the Russians, he fought the Taliban, he fought tribes, he fought warlords. All he wanted was for Afghanistan to be Afghani; didn't want anybody else in the place. They just wanted to be left alone."

As we neared Bedford, Route 460 became the 116th Infantry Regiment Memorial Highway, and that inspired both of us to pick up our pace and cut out the chitchat. I wasn't sure what to expect when the entrance to the D-Day Memorial came into view. Since the monument was maintained by a volunteer organization and received no government funding, I imagined something small, a single monument perhaps with a brass placard mounted on its base. I couldn't have been more wrong.

From the parking lot of the Bedford Visitor Center, a single lane wound a half mile up a steep hill to the memorial. As we hiked up, Overlord Arch came into view. The grand arch stood at a symbolic forty-four feet and six inches, representative of the date June 6, 1944. But after we crested the hill, we saw that this huge concrete structure was merely one small part of a sprawling display.

In his dedication speech on June 6, 2001, President George W. Bush said, "Fifty-seven years ago, America and the nations of Europe formed a bond that has never been broken. And all of us incurred a debt that can never be repaid. Today, as America dedicates our D-Day Memorial, we pray that our country will always be worthy of the courage that delivered us from evil and saved the free world."

We followed the circular parking lot's perimeter to the north side where a statue of General Eisenhower stood beneath an ornate gazebo overlooking a multi-colored flower garden in the shape of a sword. This was Reynold's Garden, a plaza that memorialized the planning that led up to the invasion. The colors of the flowers and their arrangement created the image of the shoulder patch for SHAEF: Supreme Headquarters Allied Expeditionary Force.

The next section, Gray Plaza, commemorated the invasion itself. The first portion of this plaza was circular in shape with a blue concrete floor that represented the channel crossing to France. We entered this section and followed the curved walls that ran the perimeter. Affixed to these walls were brass necrology plaques inscribed with the names of every Allied serviceman who had given his life during the invasion. I'd known before coming that 4,413 Allies had died, but as we walked the long length of this wall, I was still stunned by the seemingly endless list of names. As I read the plaques, I contemplated what triumphs these boys might have accomplished if they had not been cut down so young.

Two-thirds of the way through the circular center of this plaza, the blue concrete gave way to a channel of actual water. A landing craft brooked the boundary between these two areas, its open bay sitting mostly on the blue concrete side and the front door lying open in the water. Air cannons beneath the surface of this "Invasion Pool" caused water to spurt up in a haphazard pattern that resembled the splash of bullets from machine-gun fire. Rising out of the water were Hemmbalken—malicious German anti-amphibious obstacles made from railroad ties welded together in a crisscross pattern—and a soldier wading waist-deep through the water with his rifle carried above his head. Completing this "Ground Assault Tableau" were more statues of soldiers posed on the far shore, which was smooth brownstone to represent sand. One of these statues showed two men rushing forward and another statue showed a man crumpled to the ground, his helmet askew and his young face slack. Farther up the shore were four more soldiers climbing ropes up a rocky escarpment.

On the other side of this escarpment was the memorial's third and final plaza. Estes Plaza included the towering Overlord Arch and a statue showing the iconic symbol of fallen soldiers: a rifle planted in the ground with a helmet resting on the butt.

Quiet and reflective, we left the site and drove out to the Bedford Diner for a fairly bland meal. While we ate, we razzed each other about our respective branches until it was time to go our separate ways. Then we got serious. The necrology plaques had listed 4,413 names. That number was too close to the death toll of U.S. forces killed in the Iraq War—4,431 through the end of 2010—for me not to draw parallels. President Obama had announced his intention to remove all American troops from Iraq, but his timeline for completion kept slipping. Plus he had authorized a surge of troops in Afghanistan, a war that looked like more of a quagmire than Iraq, one that was about to suck my good friend into its depths.

"Listen, bud," I said, "Be safe over there. I don't want to read your name on a plaque any time soon."

He nodded. "No worries." Then he gave me a salute and drove off in his truck.

As I watched the taillights slip into the night, I thought again of the Bedford Boys and wondered how many of them had said something similar.

Jocelyn Corbin

Black Hat

"Are you going to vomit, cadet? Are you going to pass out, cadet? Cadet, are you alright?"

My eyes meet hers and I feel color creeping into my cheeks as the words "Ready to board, sergeant airborne," spill out of my dry mouth. I am surprised by the confidence in my voice. My stomach cramps with nauseating pain again, and my head throbs to a slow beat. I can't tell if it is nerves or dehydration. She lifts her black baseball cap from her head and reveals a leathery forehead and slicked back blonde hair that looks hard and plastic. It's perfectly tied tight behind her head. *Is she even sweating?* It must be at least ninety-five degrees in the June, Georgia sun. I imagine that I must smell so bad, but I can't tell. I'm wearing the same uniform for the third day in a row. I'm used to my own stink.

She remains quiet, returns her hat firmly back to its home, and places her hands casually on the beltline of her camouflage pants. I stand staring at her, not knowing if there is something else I'm supposed to say or do. I have felt like a failure for nearly three weeks in front of her, missing my cues, using the wrong form in my landings, and forgetting the names of equipment. She saw it all and now she wants me to fail or quit. I'm not sure yet which one. Too scared to react, I wait for her next move.

"Safety check," she says, tapping her fingers on her belt after a long moment of awkward silence. She begins walking me through a basic safety screening of equipment, which I already completed numerous times that day. It must be the one hundredth safety check this week. She tugs hard at all of my straps. I feel as though I am pulled and jerked in every direction, and I try hard not to trip as my weight shifts behind me. My hip aches from all of the falls I experienced that week and I wonder if the bruises will show all summer. I picture the colors of my hip changing over the months, like seasons in a children's flipbook. A sudden jerk at the top of my pack sends my shoulders back and my knees struggle to lock back into place. The inspection is complete. I look again to my sergeant airborne and focus on the brim of her black hat, trying to avoid her piercing eyes. I brace myself for an insult. *What would it be this time? Princess? Cadidiot?* She has many names for me. I've never felt so bullied in my 19 years of existence. *How can an adult talk to me like this? I'm trying my best.*

Her voice begins to soften and she astonishingly sounds more like a mom than a scary drill sergeant for the first time in two weeks. "What do you do if your parachute gets tangled in jump?"

Is this the same insulting beast that has been in my grill all week? I take a step back before answering. "I bicycle kick!" I say correctly, trying not to beam. Her lips turn up just the slightest in each corner, although she doesn't dare to break a real smile.

"What if it doesn't open?" she challenges back to me in a harsher tone.

"I pull the reserve!" I say with even more enthusiasm. I look down and place my hand on the large metal handle from the parachute's front casing. Doubt begins to sink in and jabs me one more time in the gut. *Am I really strong enough to pull that thing?* Droplets form at the center of my shoulder blades and trickle down to the small of my back.

"Are you ready for this?" She asks in an almost whisper, looking me directly in the eye now with such intensity, I force myself to hold my eyes to hers. Does *she* think I'm ready?

"I want to be airborne!" I declare. My hands tingle with excitement. This is it. This is everything I've been training for. It's jump day. I am the chosen one, after all, and I can't come home empty handed. I will earn my silver wings. Kent State University only gets two airborne slots for the school every year and I had earned my spot. I flew down to Ft. Benning two weeks before and was immediately thrown into this sweaty hell. Two weeks of running, pull-ups, and rolling around in the dirt have led me to this day. All that is left now are five successful jumps out of a massive military carrier—just five voluntary jumps from 1,250 feet in the air.

"Then get to it! Stay alert and stay alive." She nudges her head to the right indicating I should move to the C-141. "Remember," she warns, taking hold of my right strap, "If you don't jump, they'll push you." She lets go of my strap, turns on her heel and walks away. A flood of panic washes over me.

They'll push me? I wonder if I heard her right. I take my first heavy steps on the sandy path toward the roaring aircraft. Looking up, I see soldiers in front of me slowly hobbling up the ramp like wounded Frankenstein's monsters. *These are America's heroes?* I try hard to clear my mind and focus on the inevitable act before me. I step onto the plane and take my position. *I will not be pushed.*

The high-pitched hum of the engine drowns out all else. I strap into my seat with an oversized lap belt and metal buckle. I'm stuck now. The whole plane is a giant waiting room in the sky; all waiting for our

46

number to be called. I look around to others. I feel a little reassured that I am not the only cadet. The cadet across from me is from West Point. They have many more airborne slots than our school each year. I wonder if he feels any less pressure to succeed. His eyes are closed now and he's mouthing something. I wonder if it's steps we practiced or a prayer. *Why does he have to look so nervous?* I feel it too and reach for my barf bag, tucked conveniently under my seat. I hold it in my hands and try to read the directions on the back. The words blur together and I close my eyes tight. I breathe in, hold it, and blow out hard.

How did I even get myself into this plane in the first place? All I wanted to do was go to college. I was determined, even if it meant I was joining the Army. I truly didn't know what I was getting myself into. No one else in my family had joined the military and I knew only what I saw in movies.

"Stand up," the jumpmaster yells as he stands next to the open door of the aircraft and motions to my row.

"Stand up," my row echoes and rises in unison. This is my call to destiny.

"Hook up," he commands.

Again, we echo his words and I do my best to follow the motions of the soldier in front of me. I think of the warning my Black Hat gave me, "If you don't jump, they will push you." *Was it a warning, or was it advice? Did she say that to anyone else, or just me?*

"Check equipment."

I quickly test the tension of my straps. Thanks to my Black Hat, I am secure. I'm close enough to the door now that I spy the ground below. It looks more like a map than real land. The surrealness comforts me. *I will not be pushed.*

"Sound off for equipment check."

One by one we call back.

"Ten okay."

"Nine okay," someone confirms.

"Eight okay." I'm number seven.

"Seven okay." I strain my voice to be sure I'm heard. The plane bounces a little and my nausea begins to show its ugly head again. I inch forward, knowing my turn is about to come. *I will not be pushed. I will not be pushed.*

"Jump!" he commands my row. One after another, the jumpers disappear out the door, swept away into the blue abyss. The green light flickers next to the open door, daring me to go. My feet fight for balance as I make my way closer.

There is only one jumper ahead of me. My hand is sweaty and shaking as I hand over my zip chord to the jumpmaster. I am sure to look him straight in the eye. I'm placing my future in his hand as I commit to the moment.

Everything is happening too quickly. I need more time to get ready. My boots are close to the edge of the plane now and it's actually my turn. All of the training, all of the bruises, all of the insults have led to this moment. *I will not be pushed.*

I know from training that you don't really leap out of the door. It's more of a big step than an actual jump. It's a step I know I can take. I lift up my right boot, close my eyes, put my chin down, and lean into the wind.

The airstream sucks the breath right out of my lungs and I twist violently in the air. *Is this what's supposed to happen?* A hard yank of the ripcord heaves my body upright and I try to gather my bearings. I touch my head and grab hold of my straps. *OK, my Kevlar is still on. My parachute is open. I'm okay.* I wonder for a moment why I feel like I am being pulled up instead of falling. I soon realize it's just my parachute catching air above me and I take a deep breath. There's a sudden sense of peace and relief as I float down in slow motion. Besides the light whistling of the wind, it's quiet. I look up to the green, fluttering silk above me in a beautiful half balloon, and for the first time in weeks, I smile.

Hours later, after my fifth jump of the day, I stand on the top row of aluminum bleachers, thankful for the cool breeze. It's the only thing keeping me from leaning against the tall man next to me after the longest day of my life. I place my right hand for a moment on the metal badge on my chest to make sure it is real. I rub my thumb over the ridges of the wings and look out into the faces of the small crowd that has formed in front of us.

In our audience, it's easy to spot the Black Hats. I see my sergeant airborne, statuesque and terrifying. My hand drops to my side and I wonder why she felt compassion for me in those moments before that first flight. She wanted me to commit to the jump, and I did. Maybe her motherly instincts took over, or she saw a younger version of herself in me. She must have seen how much I wanted this. I guess I'll never know the true reason, but she helped me regain the confidence needed to take those first steps toward the plane. I would not throw up, or pass out, and I certainly wouldn't quit.

I long to shout to her, "I wasn't pushed. I jumped." I want her to be proud of me. But I know I can't yell out.

I gaze at her with gratitude and I hope in that moment that she knows how much it means to me to be in those bleachers. But her stern face does not crack, and no acknowledgement is given. She is once again just a Black Hat, but I am now a paratrooper.

Vietnam Memoir

When it comes to memory the mind is sometimes like a tourniquet releasing only as much as can be tolerated, a little bit at a time.

September 30, 1966
7:00 a.m.

Somehow I or Saint Louis University forgot to submit the necessary paper work for me to obtain a deferment. This brought back memories of May 1963 in the auditorium at Augustinian Academy when our counselor was preparing us for our senior year. He began the mundane task by urging us to study for ACT and SAT testing. Of course we sat listening half-heartedly. Then the counselor said, "Within the next two or three years some of you will be serving our country in Vietnam." We didn't pay that Vietnam message much attention, because we all knew we were going to college, and full-time student status meant a draft deferral. Going to Vietnam was never going to happen to us, hence it was never going to happen to me. Besides, Vietnam was halfway around the world. Today I am headed to the Air Force Recruiting office at the insistence of my mother, who was always farsighted when it came to matters of my future. Thanks to her, volunteering for the Air Force became the best decision I ever made, because one week into basic training, my draft notice from the Selective Service Board arrived. "Greetings, you have been drafted into the United States Army," signed by President Lyndon B. Johnson.

6:00 p.m.

Twenty of us who had volunteered for the Air Force were escorted to Union Station on Market Street where we were led to a waiting train. Each of us was shown our sleeping car. I had joined the ranks of over one million men and women who departed Union Station to serve their country. It was 7:00 p.m. when the train lurched forward and began to slowly roll out of the terminal. As I turned, looking back at the platform, I saw my mother, father, two brothers and little sister. How my family knew when the train was leaving was amazing. Yet there they were waving. Waving, as the train headed south toward the Oklahoma panhandle and into Texas. I saw cacti, desert sand, and Texas Longhorn cattle pass by taking me farther and farther away from what was familiar. This was my first trip out of the state of Missouri and only the second time riding a train.

We arrived in San Antonio, Texas, then transported to Lackland Air Force Base for basic training. On a weekend pass, I visited the Alamo to view the artifacts of Davy Crockett and the famous knife of Jim Bowie. As I stood staring history in the face, it was not lost on me the weight of the sacrifice made by those men defending our country, and I prayed, the good Catholic that I was, this chance meeting was not a premonition.

Now on to Sheppard Air Force Base in Wichita Falls, Texas, where I was trained as an Air Force Medic and then on to my permanent duty site for the next three years at Griffin Air Force Base in Rome, New York.

The Tour Begins – Twenty-three Hours to Vietnam
Sometime in August, 1969

I was on leave visiting my parents when I received a call from Sgt. Nastick: "Wesley, someone at Headquarters kicked the computer and your name came up. You are being ordered to Vietnam."

I remember hanging up the telephone and standing there for about a minute or two in silence before telling my parents I would be returning to Griffin Air Force Base because I received orders to report to Cam Ranh Bay, Vietnam.

Before deployment, I had to do what most servicemen did. I had to sell my car, a white 1962 Chevrolet Impala, and ship my personal belongings home. Now Vietnam was only twenty-three hours from McCord Air Force Base in Seattle, Washington.

September 26, 1969

I arrived at McCord Air Force Base in Seattle, Washington, a week early and waited to board the TWA flight that would transport me to Vietnam. On October 1, 1969 at 7:00 in the morning, I went to the flight line along with other military personnel and boarded a TWA plane that flew us first to the beautiful island of Hawaii, then to the Philippines for a stopover, and then forty-five minutes later Cam Ranh Bay.

October 2, 1969
1900 hours

When I landed at Cam Ranh Bay and stepped off the air conditioned plane, I walked out into the hot night air to the smell of what I came to associate with war and death, and my life was forever changed. I had begun a journey I could never have imaged. By 2200, I was transported to the 483rd hospital group where Sgt. Nastick, who had arrived a week or two earlier, issued me a helmet, M69 Flak Vest body armor and assigned me my hooch. I was walked to my hooch, found an empty cot, unpacked my duffle bag, and began my tour of duty. Around 2330, I heard an explosion (BOOM) that shook the ground and men took off running and diving behind sandbags that surrounded our hooch. I had no idea what was happening. One of the men who lived in our hooch grabbed me, slapped on my helmet, made me put on my flak vest, pulled me behind the sand bags and told me to get down . . . and stay down. That was my introduction to the war in Vietnam. I was twenty-two years old.

Just Another Day in the Nam
June, 1970
0600 hours

I was awake and it was already 80 degrees. It was going to be a hot one and I sensed it was going to be a long day as well. Needed to walk the one hundred yards to the shower room to shit, shower, and shave before going to chow. Andy and I were on the same shift, so we walked to the Chow Hall together. We got there early and ate our regular breakfast consisting of four to five scrambled eggs, five slices of bacon, three pancakes, cereal, fruit and pastry. It took us all of ten minutes to finish the meal. We sat, talked—he drank coffee and I drank orange or pineapple juice. After breakfast we returned to our hooch minus two men who have already reported for duty.

0800 hours

Andy, Dennis and I sat listening to the Temptations, Four Tops, James Brown and The Supremes on Nastick's reel-to-reel tape deck, which he purchased from Hong Kong via the military Commissary. This was part of our daily routine. We drank soda when we were scheduled to work instead of our whiskey sours and gin and tonics. The temperature has risen to 90 degrees and it isn't even 1000 hours. Drinking anything more than soda under these conditions could put us in danger of serious dehydration. Andy and I relived the basketball game we played last Sunday. Our hospital team was undefeated in our division.

Together we strategized and planned for the next game. Dennis left to report for duty, so now it is just Andy and me. We killed time waiting for the Commissary to open by rummaging through Sears and Roebuck and Montgomery Ward catalogues. We were always ordering something. That is how we stayed connected to "The World," which was everything outside of Vietnam.

0930 hours

Andy and I had already collected everyone's ration cards. It was our turn to pick up the soda, beer, and alcohol for our hooch. Servicemen were allocated ten cases of soda, ten cases of beer, and eight half gallons of hard liquor. Everyone knew you had to arrive at the Commissary on time, because if you didn't you would find yourself standing at the end of a very long line, and that meant drinking Sprite and diet soda for the rest of the month and nobody wanted that. As we were leaving, Mama San arrived to clean the hooch and gather everyone's clothes for washing. However, I was like Steve McQueen in the movie *The Sand Pebbles*, I washed my own clothes. I did however contribute to the hooch fund for other housekeeping duties.

1100 hours

Returning from the Commissary, Andy and I stopped by the mail room to check for "freedom packages" from our families. No mail today.

1210 hours

It is about 110 degrees and I am heading to the Chow Hall to eat lunch. Just another day in "The Nam."

1300 hours

Back in the hooch, I completed homework for my Abnormal Psychology Class, which I am taking through the University of Maryland Eastern Shore once a week in the evening. This is the second course I have taken and it gives me the blues. However, I need to pass this course so I can continue to build transferrable credits for Saint Louis University this coming fall. The Air Force has already authorized me for Early Out. I am ready to go.

1530 hours

At the Base Training Center, I taught English to thirteen Vietnamese who worked on base. Since spending the first four or five months studying the Vietnamese language, I had become pretty proficient at speaking and translating, and in turn became the official translator for our unit. This was especially helpful when we had South Vietnamese soldiers who had been injured in combat. I enjoyed teaching this Air Force-sanctioned class as much as they enjoyed attending and learning about America. My students arrived at 1600 hours for the one-hour class. I learned a lot teaching them, understanding their culture and most of all, they all wanted to immigrate to the United States. When they graduated from the class, the students gave me an Asia Rice hat with art work woven within. The hat was also inscribed with a prophecy that one day I would have a daughter, which in fact came true in 1978.

1700

Nothing much to do but take a shower, rest, stay out of the sun and watch *Gilligan's Island*, which broadcast at least three times a day. It was going to be a long night. I heard incoming helicopters delivering new battle causalities.

1800 hours

Finally the temperature dropped below 100 degrees. Thank God!!! Andy caught up with me and we walked across the field to the hospital. Andy has done such a great job; he has been promoted to the surgical unit working with a neurosurgeon. We headed to the Chow Hall, ate dinner, and then reported for duty.

1900 hours

I got my report from the nurses and the other medics who we were relieving from duty. We had three nurses and four medics on duty tonight. I got assigned my four patients. No one was in the recovery room section of the ward . . . yet. However, we knew patients were coming. We could feel it.

1930 hours

We did the first check of vital signs for all patients; vena cava and arterial pressures lines, measuring fluid intake, drainage from chest tubes, follies, and intestinal wounds were checked. All IV bottles were inspected and bandages were changed. If we did not get any emergencies, we should have been OK; however, we knew this was not going to happen. We had twelve hours to go, so we might as well get comfortable.

Walsh, who is one of the more senior medics, was on duty tonight and I learned a great deal from him. He and I were the unofficial team leaders for the night shift. Our first task of the night was to take care of two soldiers with extensive head injuries. They had been confined to ICU for about four days. To aid in recovery, the neurosurgeon left orders to have then sat up as often as possible. So, Walsh and I decided to try something. We would have the two soldiers face each other in order to get them talking. As they sat tied in their chairs, surprisingly they began communicating; however, their conversation appeared to us to be unsynchronized. Their questions and responses were somewhat off. After about twenty minutes, Walsh and I realized the two understood each other, but the responses were in a time delay. For example, patient one inquired, "What's your name?" Patient two would respond, "How is your day?" Patient one replied, "How are you doing?" Patient two responded, "My name is John." Their conversation continued like that for the next two or three days before they were medevaced to the Philippines for continued care.

2000 hours

The next patient Walsh and I cared for was a soldier who was burned over eighty percent of his body. We had to wash and debride him as well as check his vital signs, check his fluid output and intake and make sure he was doing well considering he was lying on the Stryker frame; however, his spirit was good. The truth was we did not expect him to live this long. Burn patients needed to be turned every two hours. Walsh and I put on the headpiece, fastened down the two safety lock handles, removed the arm boards, and disconnected the IVs. We stood at each end of the Stryker frame turning on the count of three. All of a sudden, he stopped breathing and was totally unresponsive and we couldn't check his pupils, because he was upside down. This was our first emergency for the night and we had to move quickly! We put the frame back on top of his back, fastened down the two safety lock handles, counted and turned. We placed the arm boards back on the frame and reconnected him to his IVs. The nurses were standing by ensuring we got everything right. This took all of twenty-two seconds. Our patient was comatose and totally unresponsive. The surgeon on-call arrived and did a complete physical examination. The patient's pupils were not fixed and dilated. However, after we bagged him for five minutes, he started breathing on his own. The doctor wrote an order for no more turning until he regained consciousness.

2100 hours

Once our burned patient was stabilized, we faced our next emergency. Two patients from the Army were found on the South China Beach running around naked. They were brought to the recovery ward, patient one unresponsive and patient two unconscious due to a drug overdose. Patient two was in critical condition and he was assigned to me for monitoring every fifteen minutes. Soon after being admitted, he began having grand mal seizures. His back would arch, his toes, legs and arms would hyperextend. He began sweating profusely and his heart rate was over 180 beats a minutes or in tachycardia. He kept seizing every ten to twenty minutes for the next two hours. By 2300 hours, he was dead. Patient one, his friend, who finally became responsive lying on the surgical stretcher next to him, watched him die. Patient two was toe tagged, placed in a black body bag and taken to the morgue by me. Around 0100, patient one was transferred to a medical ward for further recovery.

2300 hours

Andy came over from the surgical unit and informed us to prepare for two patients with full battle injuries. The first would be in ICU in about ten minutes. The second patient would arrive in the recovery room around 0100 hours.

2310 hours

By that time the lights had been switched off to allow the patients who were stable to get some sleep as best they could. The next patient arrived. He was an African-American soldier, like me only unlucky, who stepped on a landmine blowing off both of his arms and legs. He was conscious when he arrived to the ward and assigned to me. I checked his pulse and his heart was beating about 190 beats a minute, which was way too high. Suddenly, I could not hear his heartbeat through my stethoscope. It was like hearing an echo in a cave go silent. I thought there was something wrong with my stethoscope. The surgeon on-call was contacted immediately as we placed a cardiac board underneath the patient and initiated CPR. Fifteen minutes later the doctor pronounced him dead. I was highly upset and asked why had we stopped the CPR procedure? I will never ever forget what the surgeon said.

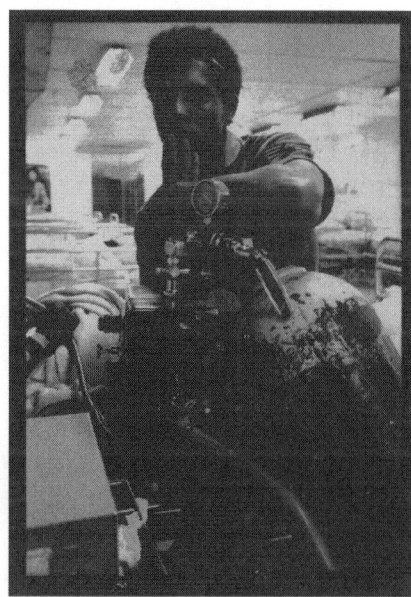

Sgt. Wesley monitoring a patient from battle causality
Photograph Courtesy of Clayvon A. Wesley

"Wesley, what type of life would he have had if he had lived?"
Once this sank in and I realized the doctor was right, I retrieved a body bag, transported him to the morgue and returned to continue providing care for the living.

0020 hours

We were too busy to go to the Chow Hall. Food was brought to us for midnight breakfast, because we keep moving, there is just no time to stop.

0130 hours

Andy delivered the next patient, a South Vietnamese soldier accompanied by his wife. He had been shot numerous times. It was my turn to monitor patients in the recovery unit. I hoped this recovery would be without challenges. I checked his vital signs every fifteen minutes at first, then every thirty minutes, and then every hour until he regained consciousness. His wife, who was pregnant, told me in Vietnamese she was happy I was providing such good care for her husband she was going to name her son Nugyen Wesley. I was honored and thanked her. By 0500 hours, he was transported to a ward for continued recovery.

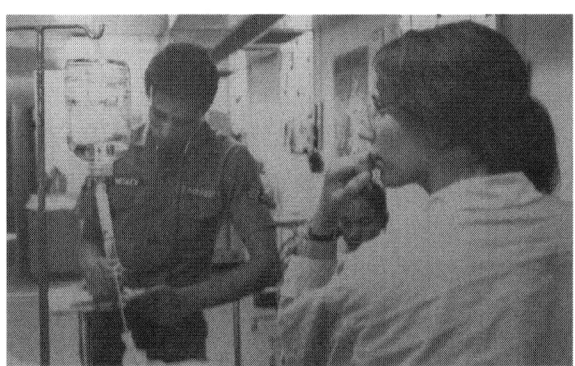

Sgt. Wesley monitoring a South Vietnaese patient that had been shot numerous times
Photograph Courtesy of Clayvon A. Wesley

0600 hours

The day shift would be relieving us shortly. We started cleaning the ICU and the recovery unit. We took all the final vital signs, checked all IVs, measured all the liquid outputs/drainages, emptied all suction devices, bathed everyone, changed their sheets and bandages, rotated everyone, mopped and swept the floor. We completed all of our dictation and submitted to the nurses for final review, approval, and filing. We were done for the night and I made it through another day.

0700 hours

The day shifts arrived. We gave our reports, prepared our patients who were going to surgery, and walked to the Chow Hall for breakfast. After eating, we returned to our hooch and picked up where we left off listening to recordings of The Temptations, Four Tops, The Supremes, and James Brown. We drank whatever we wanted, read our mail, took a shower and tried to get some sleep before it got too hot knowing that in twelve hours we would do this all over again. This was my life for eleven months and twenty-seven days.

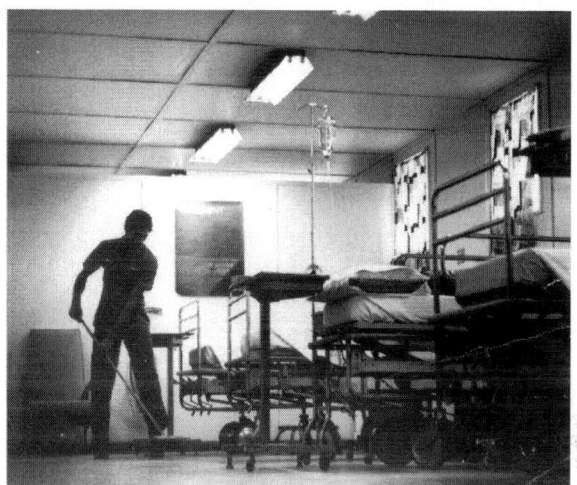

Sgt. Wesley preparing for the day shift to arrive
Photograph courtesy of Clayvon A. Wesley

Return to the "World"

In September of 1970, I returned to the mainland, arriving at McCord Air Force Base in Seattle, Washington. I was discharged from the Air Force. I flew to Rome, New York, picked up my brand new 1970 Oldsmobile four-door Cutlass Supreme and drove home to Saint Louis, Missouri. Home at last. Two days later, because I had already sent in my application for employment with letters of recommendations from the Chief Thoracic Surgeon six months earlier, I was hired by Firmin Desloge Hospital, now called Saint Louis University Hospital, as an Intensive Care Surgical Recovery Technician, and was later advanced to a Cardiac Vascular Technician caring for patients who had open-heart surgery and heart transplants. During this time, I also enrolled in classes at Saint Louis University, moved into my new air conditioned apartment, and started life all over.

For all intents and purposes, I had put Vietnam behind me. I had a career, wife, and daughter—life was good. Then an incident twenty years later brought it all back. While working for the State of Missouri Department of Social Services as a Child Abuse Hotline Investigator, David Upchurch, a colleague and friend, and I were walking from the Wainwright State Office building to the employee parking lot when I was startled by a loud noise. I remembered diving to the ground into some small boxwood shrubbery.

David asked, "What are you doing?"

I said, "Didn't you hear that noise?"

He said, "Wesley that was just a truck backfiring."

He helped me to my feet and we continued on; however, I was secretly upset he had not dived into the bushes with me. This was the first of many such incidents; thereafter, loud noises, night sweats and moments of anxiety became a challenge. Out of expressed concern from friends and other veterans, it was recommended I go to the John Cochran VA Hospital in Saint Louis, where I was interviewed by a psychologist who confirmed I suffered from Post-Traumatic Stress Disorder (PTSD). They awarded me 30% percent disability.

Epilogue

Even in the worst place under the worst conditions, miracles still happen. The burn patient and the two patients with head trauma were medevaced to Clark Air Force Base in the Philippines, and then on to Walter Reed Hospital in Washington D.C. I later heard they were recovering and had greatly improved.

The train ride to Lackland Air Force Base set me on a path to self-discovery and self-determination, away from what was familiar, and gave me a broader vision of the world. The Vietnam experience allowed me to accomplish something for the greater good. I will always consider it a privilege to have served at a time when it was not popular to do so. Through all of this and my PTSD, I tell myself I am doing well with my 70% and keep working on my 30%, which will always be with me the rest of my life. I remember the Alamo and I remember Vietnam and consider myself lucky. It's a process.

Getting On

This is the museum curator who bunkered down during the invasion and refused to leave the Ziggurat of Ur behind. He was so determined to stay he put a sign on his cinderblock hut that said, "Information," and waved to the Coalition tanks as they stormed by, kicking up dust that stung his eyes. He beamed with pride over the still visible cuneiform letters imprinted in bricks as if they were a love note from antiquity. But it wasn't just the Ziggurat keeping him from being another refugee. Next to it was Abraham's house and catacombs; he gave walking tours of both to the GIs and did not care when they held Easter mass on the steps. He liked to share a story of how Pope John Paul II meant to visit Abraham's house but that *Saddam said he'd kill him. So the Pope not come here.* He would get a faraway look that may have been hope but always sighed at the soldiers who had money he would never see, running water, cars, and cable TV Unfortunately they did not have the kind of money that could save antiquity.

*

A mother's work is never done, especially with a child who was not her son but a kid who got himself in trouble with the MP's interpreter by threatening to rape his sister. The MP thinking the boy belonged on the farm took a cow hostage saying it would not be released until the boy returned with a parent. The orphan thought the farmer's wife would do and she did, but once she learned what he said she smacked the child across the head left then right and then again, over and over until he came to understand he had years to go before he could act like a stupid man.

*

He came from Ramadi—but goes by Joe—and tended to the GI's pet, a mottled mutt kept cool in an air conditioned connex. Joe had been a big animal doctor, a vet, and now rents soft-top pick-up trucks and camel rides for a couple bucks. *This is what I do*, he says, and still manages a smile when he remembers once Bagdad had a zoo.

*

A construction contractor, whose business made all things cement for the base, was awakened before dawn. A troop had locked himself in a porta-john and bit off the end of his gun. During an attempt to rescue him the door was broken in. The investigation was swift and the body secured, but no one had cleaned up what remained. The contractor directed his crew to remove the muddied ground, wet with blood and human waste, and in its place to lay new earth, a mix of rock, dirt, and sand that he tended like a garden—his home—this land.

The Night Before

I sit on a sand-dusted cot, cleaning my rifle, in a crowded tent, surrounded by 90 or so other Marines in the Kuwaiti desert. It's just before evening chow—rumor has it we push north tomorrow, invade the country we've come here to wage war against. I am scared. Period point blank. I feel as though with the rush of excitement, I've been caught up in the wave of patriotism that's swept across the purple mountain majesty. The amount of time between me holding a remote control in the armchair of my living room while watching CNN, to sitting on this cot in this tent feels like no time at all.

As far back as I could remember, I wanted to be a Marine. My father was a Marine and he served back in the 50s. When I was a kid, he'd wake me up and carry me downstairs so we could watch old war movies together. We'd make machine guns out of our fingers and shoot each other from behind couches, and throw balled-up paper grenades at my mom and sisters. He sang marching cadences in the car and told story after story about his time in the Corps. I grew up whistling the theme from *The Bridge on the River Kwai*, and watching Jim Brown make his last-minute, always-fateful, run to freedom in *The Dirty Dozen*. Eating cold beans off a metal plate looked cool because The Duke did it; sleeping in a foxhole under the stars would be tolerable because Charles Bronson never complained when he did it on TV. I was a kid then, I'm not a kid anymore.

While I sit on this cot, while the Marines in this tent busy themselves, I think about how I got here.

*

"Uh, roger that," I said into the receiver as I hung up the phone. I sat on the edge of the bathtub. I placed the phone on the sink and leaned back against the cold white tile. A bright fluorescent light hummed on the wall in front of me. I stood, taking stock of myself in the mirror. I put the phone in my pocket and left the bathroom; I paused at the top of the darkened staircase. My daughter Emily and niece Marissa sat in the living room below. I heard cartoon laughter coming from the TV. It was a Saturday evening, the week before Christmas, and the house was bathed in the glow of slow-flashing lights. Stockings lined the bannister, decorations covered every piece of furniture. I listened to Emily and Marissa, and beyond them the sound of my mother and sisters talking in the kitchen. Quietly I moved downstairs.

"Hi uncle B."

"Hi Marissa."

"Hi Daddy."

"Turtle," I said, as I leaned in to kiss her cheek. "I love you very much."

"I love you, too, Daddy."

I tickled Marissa as I passed her and made my way through the living room. A thick, heavy snow fell outside the picture window in the dining room. Cars, already half-covered, looked like sleeping babies under their fresh white blankets. I exhaled and stepped into the kitchen. My mother held court at the far end of the table while my older sister Barbie, and my twin sister Theresa, flanked her on either side of the table.

I paused for a minute and smiled. I took a seat at the end of the table and looked at each of them. I cleared my throat.

"What did you just say?" my mother asked.

Before I could answer, the kids, laughing, skipped into the kitchen. Marissa ran to her mother, while Emily hugged me and climbed into my lap. Both started asking for more cookies.

"I said, I have to go to Iraq."

There would be no more waiting to see what our president, our military, or the country was going to do post September 11, 2001. The time was upon us to invade Iraq. It had been eleven long years since I went to boot camp, since I'd graduated from Parris Island, and now I would be put to the test in the crucible of combat. That's a long time to wait, a long time to prepare. To wonder and not know. People used to ask me if I really wanted to go to war, if I really wanted to fight. I used to liken it to someone who studies for years to be a doctor, or a lawyer; a mechanic or an athlete who practices every day for a game. Of course, I wanted to go, or rather, of course I thought I did. I had the luxury of not being this close before.

When we shipped out, I was scared I would fail as a leader of Marines. I had been a platoon sergeant at my reserve unit, which is to say, I oversaw thirty men, professionally and personally: teaching them what war might be like, everything from firing a rifle to packing the right gear. It all seemed so foreign now, so distant. We were thirty miles from Iraq and all those years of training seemed wasted. Had I taught them how to survive? Had I given them the knowledge they would need to make it home? Questions. I was sick with questions about my abilities. Maybe all soldiers headed to war are.

Back in Pittsburgh, it was the faces of the parents looking down at me with quiet panic in their eyes at the thought of never seeing their sons and daughters again that I think of now. Their looks of half-reassurance at my qualifications as their child's boss as we boarded our buses is a vice-grip of pressure I haven't stopped feeling in my gut. When we left home on that chilly January night, I pretended it was no big deal to my family. I made home videos with them. I brushed my daughter Emily's hair. My mother and sisters cried. I tried to maintain a cool exterior. An exterior that I must rebuild here every day, a façade of calmness, while war waits for us across the border. Waiting, like a summer thunderstorm on some distant horizon.

*

While I wipe and re-wipe the same spot on my rifle, I sit here and keep replaying the last few weeks in my mind. I try to organize everything we have done so far like I'm packing my gear. There is a reassurance in the packing and repacking of gear. That sense of completion that checking things off a list brings. I repack my thoughts carefully.

*

We were sent to Camp Lejeune N.C. for about a week before we left the states. We were issued more gear than one person could possibly carry, and given enough shots in the arm to qualify us as bona-fide science experiments. The anthrax pustule that has lingered unpromisingly on my shoulder, which was meant to counteract any biological weapons Ol' Saddam had in store for us, just seemed to sit there bubbly and sensitive and ominous.

"What exactly did they pump into our arms?" was the question I heard most around the barracks in North Carolina.

Our time there was filled with busywork, dental appointments, and medical screenings—I suppose the military felt it was important not to send someone into combat with a cavity or with athlete's foot. We tried not to think about what might happen to us once we got in country, once the shooting started. In the military, as in life, or at least on all those old black-and-white war movies, it was always the other guy who got "it," always the other guy who didn't make it home. I continuously looked around my unit hoping to spot those less prepared than me, hoping to find the Well-At-Least-I'm-Better-Than-He-Was kind of marine.

When our time in Carolina was over, we boarded our jet and headed to Kuwait. I mostly thought about my list of lasts; my last drive around Pittsburgh, my last look around my childhood bedroom, my last Christmas cookie, my last Will and Testament, my last everything with my daughter.

When we stepped off the last plane in Kuwait, it was like getting smacked in the mouth with an open, preheating oven. My God. The heat of Kuwait was beyond intense. My company, Military Police Company Bravo, was then bussed to a newly built tent city called Camp Bougainville, where roughly 1,000 Jarheads would call home before the invasion.

*

Our tent, built for fifty Marines, houses ninety and reeks of sweat and tobacco and gunpowder. There are cots against each wall and a row down the middle. Sea-bags and flak-jackets lie all over. We are crammed inside this tent like sardines, like inmates in an over-crowded prison. Marines tied rope across the tent and hung their trousers and t-shirts and underwear on it. Sweat socks full of water bottles pepper the tent wall. At times, it looks more like a refugee camp than military housing.

I've learned that the desert is big, and yet I can't seem to take a step without tripping over another Marine from Bravo Company. My cot rests between the tent wall and another Marine's bunk; I claimed this parcel of land because it seemed better than being sandwiched between two grown men for God knows how long. My cot is a green canvas and aluminum-framed respite from the long days on my feet. I think it was built by Oompa Loompa's because it is neither long enough to lay on without my feet dangling over, nor comfortable enough to get any real REM sleep. It is, however, the perfect height and length for slamming my shins into daily. Lying beside me, mere inches from my face, is Eddie Paluch, an outspoken kid from Buffalo. At night, we can feel each other breathe.

I sleep with my back to him, which has given the tent wall the opportunity to pound me about the face and neck when the evening winds kick up. By the third day, I became a stomach sleeper.

The camp itself houses the collective weapons, vehicles and Marines of the ever-growing fighting force. Generators hum obnoxiously while the ever-present aroma of diesel fuel permeates my nostrils.

During this time at Camp Bougainville, my unit and every other unit on base is trying to cram each minute of daylight with classes chock-filled with information we're supposed to never forget. Classes on proper desert hygiene and water consumption, periods of instruction on field-expedient trauma dressings and how to call in a medevac chopper. The frayed green notebook I carry with me always looks like a generations-old family cookbook, with rabbit-eared pages and circled entries. My indecipherable chicken scratch will be the death of me if I ever actually need to use the damn thing. Each night I return to the page entitled "Key Arabic Phrases!!" which is pretty much useless because I can barely make out the haphazard way I scribbled down "Stop!" and "Put down that rocket launcher!" underneath various pictures of the Arabic alphabet.

Without a doubt, the class on biological weapons insignia keeps me up at night the most. The all too real fear of coming face-to-face with one of those signs in my immediate future is a thought that makes me shudder. When my junior marines, beating around the bush like they do, kicking rocks and pretending to be aloof, ask if I think we will see any biological weapons, I lie and say no. I tell them to go repack their gear for the umpteenth time and clean their weapons. The fact is, I hope we don't.

The worst part of the day by far is noon, when the sun positions itself directly overhead and seems to reach down and grab me by the shoulders. At noon, being inside this tent is as close to unbearable as one might imagine. The relentless heat forces us to eat our pre-packaged meals on a hillside that runs the length of the camp. We convene behind our tent regularly for this midday chow. Our second day here, Eddie and I perch ourselves on the hillside and start in on lunch.

"What did you get?"

"The five-fingers of death, you?"

"Fucking beef stew again. What the fuck!"

As we begin, we feel the whisper of moisture being carried on the slightest of breezes. Chow seems tolerable, until I look up and see the shit-removal truck hosing down a pair of porta-johns upwind from us. The truck becomes known to us as *Pooh Bear*.

Pooh Bear and the Arabian Christopher Robin who man the behemoth truck begin to represent the dichotomy of our situation. The ever-shifting balance I face between "Keep going" and "I fucking quit." This Pooh Bear driver—clad in black, knee-high, rubber work boots, presumably swimming in feet sweat, a raggedy torn kerchief wrapped around his nose and mouth barely shielding his senses from the wretched and foul aroma wafting up from the fecal material of a thousand scared marines—his 'fuck-it I quit meter' pegs out in the red. Enter his relief, the knowledge that he will stay here, stay safe once the war begins, while every other swinging dick in this camp will be heading north to futures unknown. Suddenly the Pooh Bear driver's job seems kind of enviable. The idea that things could always be worse begins to take root in my brain-housing group. I remind myself to be thankful for what I have here, now, because noon-chow-Pooh-Bear-mist in Kuwait is still in Kuwait.

My days up till now have been like this. I wake up, get classes on various ways to kill people or avoid being killed by people, train under the unrelenting sun, eat some Middle-Eastern version of American fare in a crowded chow hall, then try to snatch a few hours of sleep between the smells of an overcrowded hooch and the snores of Eddie only inches from my face.

Nights here have been better than the days, but not by much. The drop in temperature raises spirits enough to make a difference. Night isn't the same as sleep, and sleep is a luxury I have learned to forget about. At night, my thoughts drift to our neighbors to the north. The Republican Guard. Saddam. What combat will be like. I lie awake in my cot most nights exhausted by the day's routine, the day's monotony, drifting down the river of self-doubt, my raft overflowing with second-guesses, and questions from my junior Marines, questions I can't answer. Our already jam-packed tent of men is made more crowded by our unspoken fears, which hang in the air like a heavy fog.

Some nights my platoon draws guard duty and we patrol the camp in a Humvee. I like this duty, there's a solitude to it, and it gives me time to data dump the day's events. Eddie and I usually work it together, and we take our vehicle up on a tiny hill behind the base and survey the sleeping masses below. You can see Iraq in the distance, oil wells burning in the night. The smoke creates a haze that will trap the rising sun's brilliant palette of colors; their momentary magnificence rivals any sunrise on planet earth.

Somewhere between night and day, between navigating Pooh Bear visits, between classes on nerve agents and medevac procedures, somewhere before dawn, before the sounds of angry leaders yelling at fresh-faced troops replaces the stillness, this place is calm.

No matter the time of day, the fear remains. The worry I carry for my Marines continues to gnaw at me. I need to let them learn these classes on their own, learn to take care of themselves. I can't micro-manage them. This is one of the hardest things for me to do. The time it takes showing the juniors how to do a thing, load a machine gun, dress a pretend wound, fix a broken radio, then letting them fail at it so they can remember on their own is time I don't have. The time-to-combat-clock is ticking. I want free-thinking problem-solvers, not robots, but the time it takes to mold a free-thinking problem-solver is time I don't have. Trying to replace the nervousness they feel with confidence in tasks completed is a tall order. It knots me up inside and swirls around my stomach. My only defense against their fear is the poker face I have come to master over the years. Reflecting a look somewhere between stoicism and rage, it scares the living shit out of my junior Marines. Whenever their questions start in, I must mask my complete and utter ignorance about our fates with my always-ready poker face. It takes only one Marine asking a relatively benign question for the rest to start in with an avalanche of doubt. Invariably this will occur when they have me cornered in line outside chow hall, or in the showers, or at night in the tent when they know I'm just pretending to be asleep.

"So, when are we leaving, Sergeant?"

"What do you think it will be like, Sergeant?"

"Do you think there are really a million Republican Guard up there Sergeant J?"

"How long do you think we will be here?"

"Sergeant J, will I ever have to call in air support?"

"You're going to be with us the whole time, right?"

Enter my poker face. I'll slowly light a cigarette and tell them to calm down, then redirect them to some other task. It is impossible for me to let them know I have no idea, that I am just as scared as they are, that once we cross the border they all, we all, may die. I need them to feel my confidence, soak it in and take comfort in it. I just hope I can avoid any mirrors because I'm not sure I would believe my poker face right now.

These thoughts swirl in my head while I sit on this cot and clean my rifle. Eddie walks into the tent and quickly makes his way over to me. He paces around the crowded space and stares at me pensively.

"Did you know they're serving pizza and soda tonight?"

"No, I didn't," I say, "but that can only mean one thing."

"Yeah, I know," he mumbles.

"Well, let's get some fucking pizza, I guess." And with that we walk in silence to the chow hall for our last meal in Kuwait.

After evening chow, the tent hums with restless Marines. I lie on my cot trying not to think about what is about to happen, trying to digest the cheese and pepperoni that sits in my gut like mortar-less bricks. The light from a dozen red-lens flashlights crisscross the hooch. The entire camp is in blackout, probably the entire country, and my tent like so many others is crowded around a tiny radio with the promise of a presidential address. I lie in my cot because I don't need to hear what I know is coming at first light, the invasion.

Before dawn my platoon loads up our vehicles and we drive in silence for the thirty or so miles to the border. Hundreds and hundreds of vehicles, as far as the eye can see, sit idling, the smell of diesel fuel permeates everything. We stage our vehicles and I walk the line; I nod to my Marines as I pass their vehicles. Most sleep, some watch and listen to the rat-a-tat-tat of machine gun fire in the distance. Jets and helicopters scream overhead, they drop their payloads on a little piece of land just across the border called Safwan Hill. This border town, only a few kilometers from us, is an Iraqi military observation post. We bomb it further back to the Stone Age. We drop so much firepower on the hill that the haze of smoke leaves only the slightest of outlines. It looks like the sun fell from the sky and landed on this hill. The thunderous report of American bombs and missiles seems endless.

I continue to walk and find Fitz, a young Marine from the slopes of Pittsburgh's Southside. His thick Yinzer accent calls out to me.

"Sergeant J, are you fucking seeing this? The Air Force is blowing the shit out of that place."

"Yeah, they are," is all I can muster.

Fitz had somehow managed to commandeer a satellite phone and pulls it from his pocket. He asks if I want to make a call home.

"Jesus, Fitz, how the fuck did you manage that?"

"You know me—anyway everyone else is asleep so, you want it?"

"Fuck yes!" I snatch the phone and dial furiously. After an endless number of clicks and echoes, I reach my mom. Fitz and I tradeoff making calls, I talk to my sisters and my daughter. Their voices are warm and familiar. We finish calling as many people as we can and sit atop his Humvee. We stare off into the night and listen to the explosions. Fitz and I can't seem to wrap our minds around talking to our families while standing at the border of a country we are about to invade. We decide we will keep this moment to ourselves.

I return to my vehicle and wait. Hours go by, then our radio clicks a barely readable message about moving out. I take my orders and follow the vehicle ahead of me. We pass through a giant barbed wire fence and into enemy territory. We drive at a snail's pace, maneuvering around burned-out Iraqi cars and blown-out enemy tanks. The smoke is thick and hovers above the countryside. My vehicle crew is silent. We stare blankly at the carnage we pass. Bodies line the streets as we roll across the border and drive deeper into this invaded country. Arms and legs lie about. Pockets of fire consume half-standing structures. In the distance machine gun fire slashes through the night. In the distance women and children cry.

I drive the Humvee, and simultaneously I watch myself drive the Humvee. My mind floats above me, trying to reconcile what my eyes take in. I am suddenly the Pooh Bear driver, I feel lucky, that I am sitting in this vehicle alive and not one of the dead we pass. I already miss Pooh Bear mist.

We roll through another town. The infantry engages the enemy up ahead of us. Rockets streak across the horizon, flares illuminate the early morning sky, a mosque blares the morning call to prayers.

We have crossed the border and are still alive. I flex the muscles in my feet, and in my ankles; I flex the muscles of my thighs and of my calves; I wiggle my toes. I do this for fear that it will be my last chance to do so. Along with the gear I am issued, I carry with me a measure-less weight of expectation and fear. Burdens I cannot shake. I carry a fear of landmines and sniper fire, the fear of being blown up, taken prisoner, being shot. I carry the fear that one of these things will hap-pen to one of my Marines. Each fear cycles through my mind repeat-edly, on an endless loop. I drive on, while the sun rises on my first day in-country.

Jay Harden

The Moment I Lost My Mind

It seems in some way or another, I had been at war all my life. Combat in Vietnam was just a formal interlude. This story tells how I discovered the real me, the reason I am alive today and continue to thrive.

The Saturday I met that woman, the strangest day of my life, started out badly in fear and ended up unpredictably incredible. But, after you hear my biased story, I'll allow your own conclusion.

Most of the time, we do things we like and sometimes things we don't like. But on rare days, people do things they are compelled to do against reason as if forced by the great power of the universe. That Saturday in 1997 turned out to be a combination of all three doings for me.

I started that day hopeful in so many ways and courageous in a quiet way, too—that private silence the rarer, coveted quality in me.

I had already made two decisions driving down. I resolved to return south to the cemetery, some 800 miles away, and complete my grieving at her grave exactly ten years after her untimely leaving. In my lifetime of traumas, the loss of my beloved was the greatest of all and I intended to complete my healing and release her from my heart, a great love connection, strong and timeless. I wanted to sit at her stone and talk and sing with my guitar, and find the peace I deserved that she and our infant son now had. I asked our children to meet me there, but destiny forced me to go alone. My hopes were tender enough; I was bringing a bundle of her favorite flowers, daisies. Perhaps I might see again the small yellow and black butterflies that landed one-by-one on the stone during her graveside service. They were her cherished and certain gift and goodbye to me that explained without words why she did not allow me to be present at her passing and could not die in my arms as I wanted.

With these emotions percolating in and out of my consciousness, I drove south to the land of my birth, a slow and humid place still locked in ancient ways now foreign to me.

Coincidentally, it seemed, I also resolved to confront and destroy an old unrelated fear of mine. All my life, especially in my business travels, I had been puzzled by the popularity of getting a massages. On my trips I often noticed that upscale hotels offered executive massage. Many times I peered in the door hoping to catch a glimpse, always accompanied by a slight shiver. I never dared go beyond thinking about what that experience was like. I did not consider this odd at all until a cousin told me she gave such a gift certificate to her husband as an anniversary present. Apparently, this luxury was normal behavior.

I did not know my profound fear of being touched by others originated in childhood. I only learned this late in life after war in Vietnam forced me to deal with combat trauma which revealed the extreme horror of forgotten assault at four years old, the day my father's elder twin changed my life forever, the day my childhood disappeared, the day I became wise beyond my years, the day I completely surrendered my innocent body and mind to an evil man with total power over me, the day he sexualized me before I even knew the meaning, the day I ceased normal breathing, the day I died for a time only to return, the day I assumed the blame and shame and guilt for that sorrowed experience and every one thereafter.

But this present Saturday I resolved to somehow turn and look my dragon of fear in the eye at last, before I lost my temporary courage. I did not want to go to my grave regretting any undone things, especially based on fear, just as I never regretted my loving her, knowing I would do it all again in a St. Louis second, even knowing our life together would end too soon.

I timed my trip in such a way that I could arrange my first massage enroute and, if I lost my nerve and cancelled at the last minute, none of my family would inquire about the failed nerve of a coward.

Fear is a great intensifier of distorted thought and action, and for me it was no truer than that singular day. I arrived at the appointed time and knocked on her door in a state of hyper vigilance. No one came to answer. Still I persisted, growing more uncertain of myself and the massage etiquette mysteries I did not know. My mind began to melt slowly and I convinced myself the appointment was one hour later. When I returned, the exact scenario repeated and I walked away stunned, went to the nearest trash can and crushed into it the stupid solitary rose still in my hand.

I lost control of my familiar demeanor and became enraged at myself for being so foolish, even though I could not name my error. Panic followed and I sensed my daring was leaking away. I also felt that this day of reckoning was unique, never to come again.

I drove in my strange frenzy to the nearest telephone booth at a gasoline station and ransacked the yellow pages. I was determined with a hot white burning to get this over with, off my plate, and out of my mind forever. Today was the day: now or never. Reason had departed and in its place reigned the same razor focus I used so effectively navigating a B-52 in combat and bringing all of us home safe to base 63 times, my only secret source of pride as an otherwise emotionally numb human machine killing humans.

My memory of that surreal day is hazy. Somehow I found a phone number that had choices of female voices. I went through several until I came to an elegant one that soothed me like hidden wisdom. She sounded safe in spite of a foreign accent, British I assumed.

My fear and fascination drove me to schedule our appointment two hours hence, giving me time to case the location for my safety. Then I drove to the horse farm and saw her long, winding dirt driveway past an open gate. I knew where to place my car heading downhill in case I needed to sprint out and escape.

The walk from my car to her door across a long and level gravel gap was the longest of my life. My legs hardly worked, as if drugged. I started hyperventilating, simultaneously cursing and bargaining with God, promising with great sincerity to become a monk, even a nun, if only I survived the day.

At exactly the point of no return, I noticed the black beast. A large quiet dog stared at me motionless around the corner of the house as though assessing an attack. I decided to keep going, rather than retreat. I was not coming this close only to concede to some dark pet.

The Labrador walked slowly to intercept me, a complete stranger. Even stranger, he seemed to have something in his jaws that looked like flesh, perhaps a dismembered client of hers who did not fare well. My mind was grinding, unable to make sense of the unfolding scene. It looked like a rack of ribs. But where could that come from? Not off a grill, surely. So who gives a fresh rack of ribs to a dog? Besides, any sensible dog would take his prize away to enjoy in a hidden place or bury it for later. My entire rational world was starting to crumble. Finally, he stopped and dropped the meat at my feet, then licked my hand. From out of the air, I heard a lovely voice say, "Sage, no lick." This wonder dog, destined to help my trauma healing, had made of me an unwilling friend and we walked to her together.

When I looked up, all I saw was this elegant figure framed in a golden halo of hair and a calm smile. "You must be John." Still speechless and looking around for a reality anchor, I barely nodded. She swished her full-length colorful print dress and directed me inside.

I followed downstairs to a small room (actually a former bomb shelter) lit by fragrant candles, hinting of incense. The massage table centered the room, with a daybed resting against the far wall. Soft music completed the creation of a separate world. I stammered to say that I had never done this before, hoping she would realize I needed advice. She told me to put my clothes on the chair and lay face down on the table while she went upstairs to turn off the answering machine.

When she left, my mind escaped control and I imagined cops

bursting in the door to arrest a naked man. So I just stood there. When she returned I started undressing. Embarrassed, she left the room. After I assumed the position, I heard the door open. I could not see her with my face down and feet pointed toward the door. I didn't know if anyone else was in the room. My heart raced. All I could hear was the music and the occasional ploop-ploop from pumping massage oil.

At that moment, I started trembling and wanted to be anywhere else in the world. She started chatting and I didn't know if a conversation was part of the process. My body was totally inflexible from the fear and stress. I noticed things on the wall saying she was an ordained minister and performer in the 1996 Olympics. These incongruous things, perhaps warnings, were just not adding up. I was stuck in an intolerable situation of my own creation and she was blocking my only exit. Only a miracle of some sort could save my life.

She talked of Princess Diana with casual familiarity and told me incredible stories of her life, a life of privilege completely at odds with a masseuse on a farm in Georgia. When she told me she was an aviator, one of the first female hot-air balloon pilots in England, that was my confirmation of disbelief. She didn't know I was on a B-52 crew in the Vietnam War and I wasn't about to challenge her incredible claims, not at the moment.

Her irrelevant chattering words began to blur into a continuous stream of rising and falling syllables I could no longer connect together, a poetic litany fading and flowing over me, soothing all my fears, then all my thoughts.

She, through some kind of divine guidance I never knew before and her tender personal guidance completely disarmed me, an unknown man. In that moment, for the first time in my life, my mind just quit working—the strangest, most unexpected and desired peace of my life. My brain simply filled to overflowing and gave up, unable to think through things any more. My ears drowned. I surrendered to something blissful, always sought, and unknown, and I was left as a new baby without experience or opinion, naked in the moment, revealing a magical child within a delightful man.

Her droning voice faded beyond translation and for some reason still unknown, I relaxed. My body simply yielded to her hands and voice, I suppose. I gave up trying to understand what was happening, trying to control what I clearly had no control over at all. Then I must have lost my mind, that familiar part of me that always managed to keep me alive no matter the circumstance through traumatic childhood, traumatic war, and the traumatic death of my little boy and then my beloved.

In the most ineffable moment of my life, the universe stopped and tilted on its edge in my honor. In that moment, I knew for the first time ever what it felt like to be utterly safe in the perfect present, one moment after another, gently nudged by the voice and touch of this radiant woman, this stranger I had never met. In that suspension of conventional time, I experience the unveiled connection of the universe with me in one living whole. Every possibility flowed then in a calmness born of my seeking and our meeting. I'll never be sure in words exactly what happened, but I can fairly say it was a mystical experience of finding the real and authentic me.

As I walked across her threshold to my car, now a different man as changed as a crawling critter to a butterfly, I heard a voice spontaneously emerging from inside me. It said aloud, "I don't think I'm supposed to leave yet." The words flowed out effortlessly and in my state I was no longer surprised at what I said or even her casual response. "Oh, that happens all the time. Come sit on the couch for a while." I did. She told me she was an artist and also a spiritual counselor as if these facts were an invitation to the found me.

And completely without effort or my usual shyness, I told her my story, where I was headed, every essential thing about my life, all because of the inexplicable trust we had just forged past fear in care. At that moment, my life and hers turned to a new, brighter, and clearer direction, a moment we had both waited this lifetime for, now slowly unveiling, now shared.

That moment of that day when I lost my mind started me on the path I had sought all my life but could not find or even express. Now, this moment, my stumbling ended and something better began. I accepted the avoided truth that the real me was not a deadly man but, in fact, a spiritual being.

I had found the right companion to walk with me and guide me in the process of becoming myself, another who would take my gifts in exchange for hers. In time, my destiny and the meaning of my life became unmistakable: my purpose here is to fall in love with me. Then everything else in my life will either fall away or fall in place effortlessly. In one short slice of a day, I learned that the adventure of my living is that simple and safe.

Somehow I hope to step into that house again one day when the current occupants are away and install a small inconspicuous copper plate, about four inches square, into the flooring of that tiny significant room that simply says: "This is where it happened on September 22, 1997, the moment I lost my mind."

Poetry

Poetry Winner

Michael Eaton

Kitchen Firefight

You hear the sudden warning
sounds of her staccato voice,
and don't know you've been hit
until you smell the blood flowing
down your chest—
warmer than the sun,
thicker than tears,
redder than her lips.

"Rat-a-tat-a-tat," she says
with a deadly aim and
you look for high grass,
for a shallow hole,
a burned-out building,
a dead body's shadow—
for shelter at any cost.

You walked right into
the ambush ignoring
the sound of something breaking,
of your sense of survival,
of the hair standing
up on the back of your neck—
even the acrid smell
of cooking in the air.

You'll live through this
moment of terror, however,
there were no Bouncing Bettys
waiting in the ground to de-man you,
no incendiary bombs from overhead,
no screaming mortar shells—
only machine guns firing words
without pause to cut you off
at the knees.

Randy Brown

the ground truth

Our desert watch was almost over
when charter Flight 604 banked steep
into the Anubis space
with 148 souls-on-board.
There was no moon.
The Red Sea was dark between
Herb's Beach and the coming sun.

Somewhere out there, an Infantry squad,
one of ours—after six months of "observe and report"—
was pulling a last rotation
out on Tiran Island. Back here on South Camp,
we officers slept in our trailer-park hooches,
and woke to watch the white Huey choppers
circling like gulls.
The television on our porch buzzed for days,
CNN had the story in heavy rotation,
all pictures and talk of rescue,
while we, glancing up, could see no further activity
out on the water.
Turns out, Mubarak's house was near our base.
And the British prime minister was in town.
But the possibility of a bomb was quickly dismissed.
The footprint of debris
had been too small.

Remember the '85 crash at Gander?
Or, rather, what we'd been told of it
at the memorial service a month ago?
With many of our bags already mentally packed,
we sat in the outdoor movie theater
named after the 5-Oh-Deuce,
on some cold sand near Sharm el Sheikh,
and celebrated 248 sweethearts
we didn't know
who had once suddenly failed to come home
after keeping the peace.

The truth is, we were just glad
it wasn't us.

Ryan Stovall

Desert Rain

Living spring leaves
spin down singly,

or drift in droves,
freed not by fall's frost

but snipped from boughs
by snapping bullets buzzing past.

Peaceful puffy clouds
beyond the boughs

reflect faintly
from a face

whose skin is pallid,
pouring sweat,

a sallow fallen leaf
so pinched in pain

that prayers,
spoken softly,

ask not for survival
but beg swift passage

past this fragile
failing state.

A salten rain
drops drips onto

the pale face,
and patters crater impacts

in the powdered desert dust
beside the shell

that's just become
my brother's body.

Jason Arment

Dope Tired

I still feel bone tired
when the evil hours
turn my dreams to tumult
& my thoughts to suicide

But now there is a new
kind of weariness—dope tired
the meds making me yawn
flattening me out

Not enough that my injuries
& meds cause impotence
the real insult is hemorrhaging waking hours
to grogginess & sickness

Valerie Young

Military Haiku

Formation: at Ease
Dress right dress: eyes forward March
Move as one body

Basic Training school
Smoke em' you got em': Habit
Time to dig Foxhole

Hurry up and wait
VA compensation pay
Granted or denied

Boots laced up: shiny
Uniform pressed button: clean
Pay Grade Rank Spec 4

PT Test Push UPS
Sit UPS two MILE Run: Tired
PT patch Success

22 Veteran
Suicide prevention: real
Jamieson Perry

PTSD Damn
Mental illness triggers fear
Misunderstood: ME

(Note: The bolded stanza is a dedication to my military buddy who committed suicide)

Steven Croft

Magnetic Moment

Dust blue sky above a broken egg shell
of distant mountains, time moves as
a slowly setting sun, and we seven soldiers
sit in quiet with time to wonder, how to interpret
the country spread out under us so broken by war,
while we wait for a sound of gunfire's ricochet.

A formation of birds heads out of the blue to slowly
work towards us, our mountain perch, where they'll
float by like a dream of peace, the valley quiet now,
beauty waving to us rhythmically
as dusk comes on.

Andrew Gudgel

Thank You for Your Service

They don't all count, you know.

"Thank you for your service,"
Says the fill-in nurse at the doctor's,
Who goes on to tell me
My mom must have been brave. *She*
Would never have let her son
Join the military – too dangerous.

"Thank you for your service,"
Mumbles the stretched-ear barista knocking ten percent
Off my vanilla latte. Yet he eyes the distance
Across the counter, as if I might leap over,
Demand whipped cream at knifepoint.

"Thank you for your service,"
Says the old man in the VFW hat selling red
Paper poppies by the store. We shoot the shit.
Turns out he rode a glider onto the fields of Normandy.
His eyes mist over. I shake his calloused, trembling hand.

"Thank *you* for your service," I say before I turn away.

Bruce Sydow

A Sunlit Poppy Morning

For Danella

In the wintery moment
of a late summer evening,
Vietnam chips away at my spirit
like a rasp scraping
the final fragments
of understanding.
I rise in the morning
from soaked sheets
praying not to be noticed,
but hoping to be remembered.
I want to be commemorated
in a simple Veteran's headstone,
with my patient wife
resting beside me,
tenderly cared for
on Memorial Day.
I hope the Mourning Doves
will return each spring,
and that Scouts in tan and olive green
will arrive on the last Monday of May
and reverently place American flags
into our sacred ground.

Leonard Adreon

Korea, 1951—We Paid too Much

The cold is numbing, icy pellets fall,
The air dense with fog,
Visibility impaired. The Gooks can't see us,
The sergeant shouts, "Go," we start up the hill.

Gooey mud washes over us,
We claw and slither our way up,
Elbows propelling movement, weapons in front,
Flashes of light, shells screeching, hitting, exploding.

The earth shakes, a yell, "Marine down,"
face in the muck, roll him over,
stop the bleeding, morphine for pain,
There is blood, no bleeding, no pain.

We've seen it too much, the vacant stare,
Press the eyelids closed,
don't know why, we just do it,
A fellow Marine gone forever.

We reach the top, a jumble of bodies
left behind to cart away,
Lifeless faces of Chinese youth,
The ground red from the rain washed dead.

Two hundred and ten went up the hill,
Eighty Seven reached the top,
Another hill, another day,
We own the hill. We paid too much.

Hurry Up and Wait

Sir, permission to speak, major sir. Go ahead, private.
Sir, what time is the 10 o'clock inspection, major sir?
You mean the ten hundred hours inspection, private?

Sir, yes sir, I mean ten hundred hours inspection, sir.
The inspection will be at ten hundred hours, private.
Sir, yes sir, however we've been standing here since

ten hundred hours for thirty minutes now, major sir.
It will be ten hundred hours when I say it is, private,
as he checks his watch, waits for the colonel to arrive.

Will I ever get the military out of my mind, must each
situation become another army wrinkle in time? While
I wait thirty minutes past my 10 o'clock appointment,

ponder if I should be the private, ask the receptionist
how much longer until my 10 o'clock job interview or
take the role of major and wait for the colonel to arrive.

.

Taps

Last night, I dreamed I was back in the Navy,
back even before "Don't Ask, Don't Tell,"
when American shipboard gays were patriotic
(if somewhat neurotic) surviving in living hell.

My old chief was there, left eyebrow raised
in fear as he found my boyfriend in my wallet,
his picture inscribed "Love," my chief amazed
his squared-away lieutenant might be queer.

My division crew was there, sniggering in jest
at the thought of my sleeping with another guy,
lewd comments just barely loud enough to hear
in companionways as I, shy but proud, passed by.

Word spread, things were said behind my back,
porn appeared drawn on a pipe above my bed,
but through it all I stayed true, doing my duty,
finding support in loyal shipmates, though few.

I woke this morning feeling as I have before,
glad to have served, sad my crew never tried
to fathom gays in uniform, but joyful today's
sailors sail beyond legal bigotries of yore.

Jonathan Tennis

The Longest Day

Woke up in Kuwait.

Shaved.

Ate.

Shat.

Warmed up my vehicle.

Waited for orders to move north.

Took a picture of a friend. He was dead by the end of the day.

Rolled across the berm listening to "Bombs Over Baghdad" by Out-Kast because I was from Atlanta and this war needed a theme song.

We drove.

And drove.

And drove.

We got lost. Lost in Iraq and we were supposed to liberate them.

We stopped outside of Baghdad to wait for the rest of the convoy to catch up.

I never learned to squat shit so I made a toilet out of an MRE box and left a turd in it.

I watched the sun rise over the Iraqi desert.

I watched it rise again the next day over the burning wreckage of planes left on the tarmac of the Baghdad airport.

My final sunrise over those then familiar sands would come many years later. I will never return to that place.

Michelle Brandfass

Birds of Prey

His name means Eagle.

Although known to be the symbol of the United States, this Eagle flies Recon across the far reaches of foreign distances.

It is innate for him to target, to seize, to devour and to kill. He does it all to feed his himself and his family.

An eagle lives about 20 years. I suppose, in a way, this parallels his life.

Chemicals in the air wreak havoc on him and he is a witness to other things that pose a substantial danger. His lust to fight and to survive is more fierce than that. He has learned how to successfully hunt and to be hunted.

His keen vision always keeps him aware, ready and safe.
His family lines up behind him.
He always returns home safely to the coast because he favors it so much.

He was immature when I met him and that made him dark. He is beginning to lighten as he ages and at times I barely recognize him. He is painted with pictures from his nose to his toes.

As the years pass, I notice that he has become less complicated. This allows me thoughts of a richer yet more nuanced perception of him. I observe him and will often still find his habits to be confusing. When I feel I really understand, I realize that he and his behavior are so much more complicated than my understanding.

I have recently moved inland and away from the coast and do not see him anymore.

I hear he still travels the Globe with the same steadfast determination.

I am told he still returns safely to the coast and rests along the water's edge on the Anchors of Navy vessels.

I know that he will forever soar proudly, clasping in his fist a white ribbon reading Semper Fidelis.

Fatigue

The bones in my lower legs,
and the feet with individual toes,
are conscious of the fatigue
as the sun pulls itself over the horizon.

My mustache has hardened from touching.
Between my lips and chin
is a crease of oil
which contains the smell
of everything I touched today.

The mouth I pout with
is a crust of coffee and cigarette stain
with acid corners that make smiles
like a film fade-out.

The hard part is over.
My eyes, like a runner's second wind,
have ceased to sting,
and even the far tent walls
begin to come in focus.

A yawn cracks my face,
and the deviated septum
clicks like a meter in my nose
until I notice it,
and trying to count, it stops.

In the lower back a muscle vibrates.
They say it is a trait in our family,
a roaming tick which plays a tune
on our fibrous parts.

I wish there was no need for sleep,
that the mind never required rest,
or that it could turn itself off
when I find a bed too early to find sleep,
or that it could recognize this morning
and refuse to believe how fatigued I really am.

Then reveille is announcned,
and duty calls me
back to my life
as a soldier.

Ben White

Recordings for Later Listening
Lost Rounds

After Tim Schumacher

There was no alcohol –
               Leadership was easier without it –
So after I was discharged
You might think
I was just making up
               For lost rounds

But that's not it –
Alcoholic drinks are just better for me
               They provide more clarity
Especially over sports drinks –

               There were always sports drinks –
               Powerades and Gatorades –
               To fight the heat,

But when I taste
Or even smell one of those
A desert wind blows sand across my mind,
My stomach explodes with roadside anxiety,
And I can't even
               Hold the bottle steady.

Combat Landing

The pilot seeks to demonstrate the six degrees of freedom and narrates

Yaw, pitch, now, ROLL
                Just enough to scare the newbies in the back
Hold onto your guns and guts —

This is the rollercoaster old-timers say saves lives from mortars and
outside the aircraft releases flares in a crescendo of flashing light as the
warbird makes a hard landing.
The human freight shuffle about in ballistic vests and helmets as some
make their way down the back end of the C-130 and a few trying to
avoid the fresh vomit splatter across the floor exit out a side door just as
the base emergency system screeches,

"INCOMING, INCOMING, INCOMING."

The new arrivals, weighed down by packs, run like turtles on amphet-
amines to the nearest bunker. Inside was the flight line safety guy
wearing Levi's, reflective vest, and hard hat. Seeing the group of clean
uniforms, except for the one with a stained crotch, he laughs, "Welcome to Iraq."

Jennifer Duff

Lonesome Rising
After The Great Santini

After the stormy dusk, live oak roots seep
from supple cordgrass into rusty metal tombs,
Beaufort's drowsy Sound swallowing
a grand army of Black men,
valor displayed like shadowboxes
on weathered Yankee coats as battle cries echo
Great Blue Herons,
and skeletal remains rejoice under the sandy soil
of hollowed charnel graves.

During night's intruding shade,
when cotton rats scatter through centipede fields,
reinterred quartermasters rise above the misty
southern fog, singing buoyant brews like John Brown's
March while feasting on a plethora of Frogman's stew,
polished moon shining on the lofty granite obelisk.
Amid the wire trappings of thorny thought,
war whoops from Fort Morgan and Honey Hill wash
over lawn level markers as Red-Winged Blackbirds
revel in quiet quests before the sun takes its place
to blaze on floating caskets wafting on
the soiled Coosaw River.

As light peeks over husky hangings
from creamy Spanish Moss, gray ghosts, slinging
dreggy Rifle Muskets, retreat behind yesterday's
saintly crusade. Before soupy vapors pester
morning's virgin call, fathers from the 55th Regiment
rally round the flag while unbraided promises push
the costal breeze through limestone piers
of inky iron gates, affinity basking
in the glimmer of Egrets sailing
above the salty Spartina marsh.

Jeremy H. Warneke

Memorial Day

I remember that
in the mid to late afternoon,
it was warm;
I was sleeping.
I had my earplugs in,
the orange rubber,
heavy-duty kind.
I needed something
to drown out the noise
of carpentry work,
the banging, the hammering,
the assembling of plywood,
soon silenced by
a gunshot.

I remember that
in the confusion to which I awoke,
men scurried about.
A woman, his friend,
screamed and cried.
Others rushed the body out
on a green or brown woolen
Army blanket.
His blood trailed
in dark red drops,
smacking against the smooth
glossy, concrete floor.

I remember moments later:
my whole company, ordered
outside into the courtyard,
waiting an hour or more for
the inevitable,
while those who knew him
moaned, cursed and reflected upon
a man, just a boy really,
who had shot himself.

The Shirt

There's a shirt on a hangar
clinging to the doorsill
of my walk-in closet.
I bought it for $7.50
at a thrift store in San Diego
the other day—all wrinkled.
Blinding red, flared collar,
white frangipani pattern—
I would pack it for Hawaii,
but not a warzone.

Maybe I'm a fatalist—
but I'm going to Afghanistan
and don't even know
what my full role will be—
I'm a Navy corpsman
under an Army billet
fulfilling some enlisted
medical role for NATO.
I arrive and process
through Kandahar.

The shirt will await,
hanging in plastic protection
in my darkened room,
windows shut, door locked,
dust gathering on bookcases,
maps and photos—
a voice in the walls,
a life in the gray
waiting for the door
to unlock, open.

Photography

Photography Winner

Breanne M. Pye

Patrolling the Arghandab

Bill Glose

Desert March

Reunited

A. Sean Taylor

Night Land Nav

A. Sean Taylor

Weapons Qual

Jay Harden

Lethal Beauty

Breanne M. Pye

Rafi's Miracle

Patricia Joynes

A Town Registers Veterans for Discounts

Michele Davis

Coronado Bridge, San Diego

Interviews

Nancy Brewka-Clark

Haven in the Jungle

I learned soon after Tom came to work at the daily newspaper where I was employed in the early 1970s that he was a Vietnam veteran. The war was still dragging on, but he'd distanced himself enough from it so that no one could tell if it bothered him when the stories came rattling in over the newswires. Usually they were accompanied by grim pictures of smoldering villages, camouflaged soldiers and tanks rolling through wasteland created by bombing and Agent Orange.

Ironically, when he was drafted in 1965 his parents had to look the place up on a map and came away satisfied that he'd be "safe" in such a distant outpost of the world. His deployment seemed almost casual, his trip to Saigon accomplished on a commercial airline chartered by the Army. After ten days in Tent City, which was exactly as glamorous as it sounds, he was assigned to the 13th Aviation Battalion in Can Tho some eighty miles to the south as an administrative aide to a helicopter unit.

A former French luxury hotel stripped of its trappings served as his barracks. Instead of jardinières overflowing with orange, yellow and saffron chrysanthemums, the doorway was flanked by barrels filled with sand where he fired his rifle at the end of the day until it was empty of live ammunition. Eventually, he'd be firing at the enemy but in those early days it still seemed more like a dream than a nightmare for a Rhode Island boy to be stationed in the tropics.

We were already married when I first saw the slides. "Who is that?" I asked, studying the image being projected onto the screen in our living room.

"Miss Snow," Tom said, "the unit secretary."

"She's beautiful." The words hung in the air as I tried not to be jealous of this young woman, delicate as a porcelain doll, he'd known so long ago. "What a wonderful dress. Did all the women dress like that?"

Tom looked confused. "There weren't that many women." He thought for a long moment before adding, "Her outfit—it's called an Ao Dai. It isn't a dress, really. It's kind of a long silk tunic over white pants."

"Did you know her well?"

He shook his head. "Not at all."

"So, she wasn't married?"

"I guess not. We were told to call her Miss Snow but I don't really know what her native name was." He clicked the hand-held control and another slide came on the screen. "This is what I wanted to show you."

He'd taken the shot from a window high above an open courtyard. In the background a forest of native hopea trees still flourished. I knew from a cousin who was deployed to the Mekong Delta three years after Tom that half the country's trees would fall victim to a massive land-clearing effort by U.S. forces designed to eliminate guerilla attacks, just as bombing would reshape the Mekong River itself. Ancient bricks formed a mossy floor. Like the stately hardwood trees, the walls looked as if they'd been standing for centuries. He'd caught someone in mid-stride, a young bespectacled man with a shaven head, his slender body draped in an orange robe. In the corner of the photo I could just see a pair of sandaled feet and the hem of another robe. "Buddhist monks?"

"I taught them English for six months." Tom smiled at the memory. "In French."

"I didn't know you knew French."

He shrugged. "*Un petit peu.*"

Even I knew that meant "just a little bit." "How'd that go?"

He laughed. "They cheated like crazy."

"Oh, come on."

"Really." He laughed again. "During tests, I had to split them up."

I tried to imagine those sober young men bound to another faith so foreign to mine deliberately violating one of western education's most strictly enforced rules. "You mean you'd catch them passing notes or eyeballing each other's paper?"

"No, no. They'd actually lean over to share the answers out loud. And when they didn't know, they'd literally put their heads together and talk it all over. They were shameless!" He grew sober. "Eventually I figured it out. They'd been trained from birth to help each other out. In their minds it wasn't cheating at all. But it was my job to see that they did it our way."

"What do you suppose happened to them all?"

His face remained calm but his voice caught. "I have no idea."

I imagined the images frozen on the screen coming to life. The young monks would step forth to speak in the perfect English they'd never learned from a young Army draftee with the grand title Specialist Fourth Class to tell him they'd all survived and flourished. And the lovely and fragile Miss Snow would crank a sheet of paper in her typewriter to tap out her memoir of a time when the devastation of a long, cruel war seemed impossible and nothing would ever change.

Billie Holladay Skelley
Vietnam Vignettes: The Trauma and the Therapy

*Three years ago, I met Ronald Carl Mosbaugh, a Vietnam War vet-
eran, in a writing group where we were both trying to improve
our writing skills. Since that time, as we have worked together on
essays, stories, and articles, I have gleaned bits of insight into the
trauma he experienced in Vietnam. I also have witnessed how writing has been therapeutic for his post-traumatic stress disorder (PTSD).
What follows is a composite interview based on those interactions.*

The Trauma:

I was trained as a Navy Corpsman and attached to the First Marine
Division, Second Battalion, Hotel Company. I was in Vietnam for thirteen months during 1966 and 1967, and during my tour, I received the
Silver Star, Bronze Star with a Combat "V," and two Purple Hearts.
When I arrived in Da Nang, I was only twenty years old and very innocent and naïve. I guess I feel like my youth was stolen from me by the
things I experienced there. I witnessed so many gruesome atrocities
and scenes of incomprehensible horror—it changes you. You can't help
but be changed because they were things no one of any age should see.
Things you cannot forget.

Being a corpsman in Vietnam was a dangerous job. Every time
Marines went on patrol or on a mission, a corpsman had to be with
them. It was my job to render medical aid when a soldier was injured
or wounded. It didn't matter where the soldier was located—in open
ground, in the middle of a battlefield, or submerged in a rice paddy.
The call that would come was "Corpsman Up!" Whenever I heard that
call, I knew someone needed medical attention and I had to go. They
were depending on me, and I went, but I can still hear the rifles firing
around me, the grenades exploding close by, and the helicopters flying
over my head.

Corpsmen knew they were putting their lives on the line each time
they went out. Even though other Marines would provide fire support
while you were running toward a wounded soldier, you knew every step
you took could be your last. I remember so many times fearing for my
life. You never grow accustomed to the high-pitched sound of bullets
whizzing by your head or mortars and grenades exploding near your

feet. It was a suicidal waltz known as "doing your duty," but I did it. Once, I remember running toward a wounded Marine, and one of my canteens was hit by a bullet. In the melee and pandemonium, I thought the liquid that ran down my leg was blood, only to find out later it was water! I remember feeling both relieved and terrified at the same time.

To make matters worse, the Viet Cong had a bounty on corpsmen. They knew if they could eliminate a corpsman, more Marines would suffer and die without the benefit of medical care. If they could put a corpsman out of commission, it would put other Marines out of commission, too. So, every time you ran onto a battlefield to provide aid, you knew the enemy hoped to take you out first. Consequently, because of the bounty and the nature of the job, many corpsmen in Vietnam were killed in action.

I remember one particularly traumatic day well. I was treating a wounded Marine in a rice paddy, and while I was treating him, he got shot again. I have no doubt that the bullet he took that day was intended for me.

Another time, a Marine and I were taking refuge from sniper fire behind a stack of sandbags. We were talking, and when the shots got close, we ducked down. When I raised my head a little to see what was happening, I realized he had been shot in the face and was dead. I think that bullet was also meant for me instead of him.

Being a corpsman was also stressful because difficult and disconcerting decisions always had to be made. I might be treating one Marine and another call would come for "Corpsman Up!" Then, I had to decide whether to stay with the first injured soldier or leave him to go to the new injured soldier. I tried to triage as much as possible, but often there was no way to know how serious the next man's injuries were. Common sense made it clear I could be in only one place at a time, but which place was I needed most? I also had to decide where my limited supplies, especially pain medication, would be of best use. I hated to deny anyone who was suffering, but I could carry only so much pain medication, and it had to be dispensed where it was most needed and effective. Since multiple injuries were the norm on most missions, it was always troubling whether I had made the right decisions.

When I prepared to go on a patrol or a mission, I always tried to load up on my medical supplies—so I could be as prepared as possible. These supplies included such things as battle dressings, ACE wraps, tourniquets, ointments, scissors, salt tablets, sulfate patches, morphine syrettes, and casualty cards. In addition, I often took cigarette cellophane wrappers to plug chest wounds and a fifty-caliber round for twisting tourniquets to stop bleeding. I often felt apprehensive, as I prepared my supplies, because I knew for many of the men it could be the day their lives would be changed forever by some terrible injury or it might be the last day they would be alive.

On one mission in September of 1966, near a village called Vinh Hoa, we were ambushed and overrun by the enemy. When the "Corpsman Up!" call came from the First Platoon, I started running across a rice paddy toward the fighting—a distance of about seventy-five yards. The suction from the muddy water in the rice paddy made it difficult for me to lift my feet. I felt like I was moving in slow motion. As I ran, bullets splashed in the water around me. By the time I reached the injured Marines, I was terrified and exhausted. I tried to focus and perform triage, but it was extremely difficult while the hand-to-hand combat was going on all around me. The fighting was bloody, loud, and intense. I remember the noise from the weapons firing and the grenades exploding. It was deafening, and yet, I could hear numerous Marines screaming and moaning due to their injuries. Everywhere I looked, there were bodies scattered across the ground. I felt overwhelmed. I needed help, but there was none. I did the best I could, but it was pure chaos. We began that patrol with ninety Marines, but after the battle, only twenty-six of us walked away from the slaughter.

On another mission, we walked into a village that had been hit by an air strike. The place was filled with death—civilians, animals, everything. It was heartbreaking and disturbing to see young children as casualties. You felt so out of control. Your mind couldn't get around the degree of destruction and the amount of loss. It was beyond comprehension—because everything was dead—people, pigs, chickens, water buffalos, everything. We had to bury our emotions just to complete the mission. We made an effort not to think. You tried to function on automatic. Otherwise, you'd be overwhelmed and not function at all. Feeling empty and being numb became a way of life for me. There was just so much turmoil and devastation. You had to soak it up and go on because there was nothing else you could do.

So many similar patrols and missions still circulate and swirl in my brain. Sometimes my memories are distinct, clear streams, but other times they run together like muddy floodwaters. I feel like I live under the cloud of these memories.

I especially remember the faces of several of the Marines I was called upon to treat. One of my vivid memories involves a Marine who fell and landed in a pit filled with sharpened punji stakes (bamboo spikes). His injuries were terrible. I also remember other Marines who were splattered with white phosphorous and suffered extremely painful burns. I often recall the face of a man who stepped on a land mine and was blown to bits. I remember this Marine well because I was only a few feet from him, but I got hit with only a small piece of shrapnel. I remember another Marine stepping on a land mine and screaming in pain. When I got to him, I saw that both of his legs were barely attached. He was bleeding profusely, so I applied a tourniquet to each leg and put battle dressings on his other wounds. I gave him two ampoules of morphine for his pain, but there was not much more I could do. He went into hypovolemic shock, his breathing slowed, and he died with his head in my lap. Another Marine, shot in the head, died before I could do much of anything. At times when I was caring for these Marines, I felt so helpless. I often felt enormous guilt that I could not do more to save them.

Even now, I can see the ashen color of another Marine's face who was about to die. I remember the cries of a few men who wished they would die because of the seriousness of their injuries. I can hear men pleading for pain medicine to ease their suffering. I can even remember the last words of a few soldiers who knew they were going to die. I could hear the terror in their voices as they realized they were not going to make it. I treated hundreds of wounded soldiers, but hearing the terror in a dying man's voice stays with you.

Besides the enemy in Vietnam, you also battled the elements—especially the rain during the monsoon season. The rain there had a heaviness to it. If you held your helmet out in those terrific storms, it would fill in just a few minutes. I did not enjoy running patrols in that rain. Sometimes you couldn't see ten feet in front of you, and that was always worrisome. The wet weather also made it difficult for men to maneuver quickly, problematic for calling air support because of the low clouds, and harder to arrange for a medevac. Often for medevacs, I just had to wait for the weather to clear. The rain also caused low areas to fill quickly with water, forcing snakes to move to higher ground. There are numerous venomous snakes in Vietnam, and snakebites were always a concern. The rain also seemed to encourage the mosquitoes, leeches, and ticks to come out in force.

I remember on one patrol, it was raining so hard it was difficult to walk. As it started to get dark, we made camp, covering ourselves with our ponchos. When I awoke the next morning, I heard a Marine screaming for help. When I got to him, I saw he was covered in huge, blood-engorged leeches. They were all over his body. I used Zippo lighters to get the leeches to fall off. When the fire hit them, they released, but it was a lesson for all of us to always check for the slimy creatures. Another Marine got a leech embedded in his ear. It was so deep and attached, I couldn't get it out. I ended up having him evacuated by medevac to a Battalion Aid Station.

Beyond battling the snakes, insects, and other creatures, I felt like I was a "foot doctor" during much of my time in Vietnam. I saw a great deal of trench foot, or immersion foot, because of the wet weather. This occurs when the feet remain wet for extended periods. It may not sound serious, but trench foot can lead to significant bacterial infections with extensive swelling of the feet and ankles. I saw feet so engorged that Marines could not get them into their boots. When I had any extra time, I was always inspecting feet for blisters, cuts, infections, and other problems. I handed out dry powder and antibiotics, and I encouraged everyone to use dry socks and to expose their feet to sunlight when possible.

I also assumed the role of chaplain or counselor when necessary. Marines came to me for their physical problems, so I guess they assumed I could handle other types of problems as well. At any rate, I seemed to always be dealing with some form of homesickness, stress, anxiety, or depression. I wrote letters for soldiers and offered prayers for those who wanted them. Many times, I remember holding injured Marines in my arms and trying to pray with them as they died. Some would ask me, with their last breaths, to write to their mom or girlfriend. Usually, they wanted me to tell their family or their sweetheart that they loved them. I tried to provide comfort and lend a sympathetic ear, but I often felt like I was riding an emotional rollercoaster. I still feel like I'm on that ride at times—especially when I have flashbacks or nightmares.

One Vietnam nightmare has proved particularly haunting for me. It manifests as a recurring dream where I am treating a Marine with a bad gut wound who is bleeding out. He is in great pain, and I am focused on trying to help him. Suddenly, I look up and see an enemy soldier staring at me. He is only twenty feet away, and he has a rifle in his hand with the barrel aimed in my direction. My hands have been busy treating the Marine's injuries, and I realize my .45-caliber pistol is still in its holster. My rifle is lying a bit away on the ground. This enemy is about my age, and his dark eyes are looking right through me. I'm tremendously scared. I can barely breathe. Time seems to stop. I know this adversary could kill me easily at any second. My life is in his hands. I wait, frozen in place, expecting death—but he doesn't shoot me. The injured Marine groans in pain, and I turn my attention back to him. When I look back, the enemy soldier is gone. I never stop wondering why this man didn't kill me.

My thirteen-month tour in Vietnam seemed like a lifetime. It was like living in Hell with horror and mayhem for roommates. Every day I was there, it felt like I was on Satan's playground, dancing with death.

I think it all just added up—the stress, anxiety, wounds, burns, blindness, loss of hearing, loss of limbs, disfigurements, and deaths. The wounded and killed I saw were often between eighteen and twenty-two years of age—so near my age that their loss really hit home. There was just so much physical, psychological, and emotional trauma every day. It was overwhelming. In addition, I was constantly dealing with fatigue and "survivor's guilt." Why did I live and others died?

I don't know. Only God can answer that question.

I just remember feeling so grateful to be able to go home in August of 1967. I thought if I could just get home, everything would be better—everything would be normal again. I didn't realize it then, but things would never be the same. I was bringing demons home in my duffle bag that wanted to stay with me. In my head, heart, and soul, I was carrying thirteen months of combat wounds and battle scars home. I didn't know it, but I had been wounded emotionally for life.

It didn't help that our homecoming was hardly welcoming. Back in the States, when I was discharged from active duty, I was advised not to mention that I was returning from Vietnam, and I was told not to wear my military uniform. Because of the riots and protests near Camp Pendleton and around the country, we were all instructed to wear civilian clothes and to just leave.

"Just leave." The military is good at training you to be a soldier, but not so good at training you how not to be one.

We left through the main gate on a bus, and I remember the bus being hit almost immediately by protesters throwing eggs, tomatoes, and trash. Their yells and jeers were not welcoming in the least.

When I reached my hometown, it was better, but not great. I remember, a few months after I arrived home, there was to be a holiday parade in our town. Our local Naval Reserve Center planned to participate. There were about three hundred reservists involved, and some of the men were to carry flags and others rifles. I was given the honor of riding in a convertible with a banner on the side that stated "Silver Star Recipient." We assumed it would be a patriotic and festive affair, but we were wrong. The crowd yelled in protest and offered several insults. The boos and jeering were bad, but at one point, a woman ran toward me in the car and started hitting me with her purse!

I was home, but it was quite obvious that the world at home was very different from when I left.

In time, I came to realize that I was different, too. I felt like an outsider. It seemed like I was just visiting and I really belonged somewhere else. Some soldiers say they were "battle hardened" in Vietnam, and I guess that was true with me, but I mostly felt numb and empty. I felt like I couldn't feel much of anything. My self-esteem was low, and I felt like a failure.

The Therapy:

For almost fifty years, I refused to talk about my Vietnam experiences—even with my family members. I suffered in silence with flashbacks, nightmares, panic attacks, and bouts of depression. I didn't think anyone could understand what I was going through. I felt detached from people and society. For almost half a century, I faced my demons alone.

My wounds were not physical or visible. They were inside me. I think experiencing life-threatening situations daily for so long was just too much for my brain to process. Instead, I trapped all the sights, sounds, and smells of Vietnam in my head and stored them. These memories became so deeply imbedded within my brain, it was easy to relive them over and over. I created a time warp, and I was stuck in it.

On some level, I knew I had PTSD, but I didn't want to address it. To give my problem a name made it more real. I just wanted to stop reliving events in my head. I wanted the memories to go away, but they didn't.

Finally, when my PTSD became so severe that it was interfering with my work and family life, I sought treatment. I became an inpatient at the VA Medical Facility in Topeka, Kansas, where I attended a seven-week, intensive PTSD treatment program. I lived at the facility during this time.

One of the first things they asked me to do in Topeka was to write about my "trauma" experiences. To say I resisted their suggestion is an understatement. I couldn't conceive how they expected me to write about thirteen months of terror. Where would I start? Where could I start? It all ran together. I just couldn't understand why they wanted me to write down things on paper that I hadn't talked about with other people or even admitted to myself.

My hesitation and reluctance diminished when one of the psychologists with the program told me I had to follow their rules or I might as well go home. I knew I could no longer face my demons alone. I was past that point. I needed help. I couldn't go back home, so I finally picked up a pen.

It was not an easy task. Confronting demons never is, but when I came to understand that writing was a required part of the therapy program, I tried. I tried really hard.

After much effort, I wrote my first story.

Then I wrote another. Then another. It was like a dam had been broken and suddenly I couldn't keep the memories from spilling out.

Through the program, I finally realized that the only way to heal my wounds was to expose them for all to see. Too long they had been hidden in my heart, mind, and soul. If I was ever to be relieved of their weight, I had to reveal them and share them.

Writing became a healing mechanism for me. It helped me become more aware of the triggers, certain sights and sounds that make me relive events. It has helped me to process my trauma. It has enabled me to compartmentalize some of the demons so they are easier to handle. For me, putting my thoughts on paper has been cathartic. It has reduced some of the enormous weight I felt.

All told, I have written more than thirty stories about my time in Vietnam, and I have published a book entitled *Marine Down, Corpsman Up!* about this part of my life. It has not been easy for me to do this. Many times, the tears flowed. At times, it was quite painful to remember. With each effort, I felt like I was opening up an old wound. It was difficult, but I did it, and I can now say I am glad I did it.

My writing has changed my perspective on life. I think it has helped me to grow and to become a better person. I know it has helped my negative outlook and low self-esteem. It has helped me to heal. I appreciate each day more now, and I don't take God's blessings for granted. I am more grateful for what I have. I feel like my writing has given me another chance to make peace with myself, with my life, and with my God.

I also have given talks to groups of veterans about my writing. I tell them about my efforts and how putting my thoughts on paper has helped me with my PTSD. I share my stories with them, and they have been well received. This has increased my self-esteem and improved my outlook. I hope by sharing my stories it will help other veterans who are suffering. I tell them it isn't easy, but you can live with PTSD.

For me, Vietnam was Hell on Earth, but it has made me who I am. If anything, it has strengthened my faith. You can never get your life back the way it was, but if you can accept the damage and the breakage, you can work to build yourself a new life—one that is hopefully stronger, more resilient, and better.

Writing and talking about my combat experiences has helped me, but the invisible wounds of war never completely go away. I still suffer with PTSD, and I still attend PTSD group therapy sessions in Mt. Vernon, Missouri. When you spend thirteen months wondering each day if it will be your last one on the planet, it takes its toll. You know death could come at any second, and that fear becomes ingrained. It gets so embedded, it can keep attacking you. All I have to do is close my eyes, and I can be right back on the front lines in Vietnam. In a split second, I'm a corpsman again. I can see grass huts burning, water splashing in rice paddies, and blood flowing from gaping wounds. I can smell burnt bodies, decaying flesh, and the pungent odor of explosives. I can hear the whop, whop, whop of helicopter blades whirring, the rat-a-tat-tat of M16s firing, and above it all, the woeful cries of wounded Marines calling for me. The trauma never totally leaves. It always haunts me, but I keep trying to face it, accept it, and move forward.

Rod Martinez

Raymond's Journey

World War II vets are literally a dying breed. Not only because of age but because they are considered the greatest generation ever to have lived. WWII Vets didn't come back and share the bravado of whatever happened over there. They came back home, got jobs, and returned to civilian life as best as they could. Some never talked about it, some only barely. Among those great warriors was Raymond Valdez, Jr.

Raymond enlisted in 1941 at age seventeen.

"My brothers decided to enlist and I knew it was something we had to do," he says.

Raymond joined the Navy, his two brothers separately enlisted in the Coast Guard and the Army. Already mechanically inclined, the Cuban-American born in Tampa, Florida, was assigned as an airplane mechanic but also ran test pilot duties when needed. The stories he's shared with family have entertained and brought deep thought—like the time he was so tired that he decided to take a nap on a table and was awakened by his commanding officer who didn't think the stunt was funny at all. Valdez was sent to his barracks to suit up, and put on his bag and walked outside in knee-deep snow. Coming from the Sunshine State, that was nothing but punishment to the man who stood a full five feet tall. It wasn't until a senior officer saw him outside and asked what he was doing till he got pulled out and got his commanding officer in trouble. He loves telling that story.

Raymond eventually served as harbor patrol at the entrances to the Panama Canal during WWII which was the main route for the transport of U.S. troops and supplies from the East Coast to the Pacific. It had become known to US intelligence that the enemy made plans to send subs through the canal so the locks were under constant protection against sabotage. This job became part of the United States Navy. The Air Corps forces in Panama worked with the Navy during this period in the Canal Zone.

"The fact that I was bilingual made me a perfect candidate for the job and I loved it there." He smiles.

Raymond lost his left eye due to a mishap while working on a plane engine but it never hampered his spirit. After the war was over and the men came back, he opened a garage shop off Lake and 15th in historic Ybor City, Tampa, where he worked as a mechanic. He married Cuban immigrant Aida Gomez, who escaped Castro's regime, and they had two daughters.

"My dad was and is always there for his daughters," says his youngest daughter Nancy. "He would give the shirt off his back with no hesitation, no questions asked. This is why customers always wanted their cars fixed for little or nothing. Now in this day and age just lifting your hood costs money and my dad did it for free. My dad would have a BBQ at his garage every Saturday—opened early and closed at noon, but not before he fed all his customers, from the mailman to his neighbors surrounding the garage, sausage hot dogs."

After almost fifty-five years together, his wife succumbed to illness and passed away in January 2016. Raymond will never forget the night his wife passed away. He was asleep and felt her presence: "She came over to the bed and kissed me goodbye."

Aida couldn't get out of her bed, but he felt her come to him to bid farewell.

Raymond is the proud grandfather of three grown grandkids, aged twenty-one to twenty-nine.

"Raymond is a hero to me," says Rod, his son-in-law. "My father died in the army when I was two during a Vietnam training mission, so I have always looked up to the military for their sacrifices. I was Raymond's volunteer on the Honor Flight in 2016 and it was an experience he will never forget. Obviously, the people who run Honor Flight understand the mind-set of the aged veteran and I commend them and their project for what they do. They serve with a dignity I have never seen before. It was an awesome experience for Dad."

During the annual July 4th parade in the Tampa area, his youngest daughter and son-in-law noticed something that had never caught their eye before.

"We got there early so Dad could get a good seat. Every time military personnel would march by, Raymond would stand erect and salute. It got to the point that every time he did that, the soldier or officer would stop, salute him, and then shake his hand. He was wearing his Navy Vet hat. They would thank him for his service. It brought tears to my eyes. At a parade where we were honoring the patriotism of our wonderful country and the men and women who served (and died) to keep us free, these men and women would stop and take the time to thank him. A dad even walked up to him with his little boy and said, "Say thank you to the hero, son."

"I knew right then and there that somehow I had to get him in the parade for the next year and I spoke with the coordinator and she was elated to have him for the next parade," Rod said.

Raymond rode in style, sitting in a convertible with a huge sign on either side, "Raymond Valdez, WWII Navy Veteran," for all to see. He was a rock star that day and was the subject of a newspaper article because of it. Soon the Veterans' Day parade given by Tampa's VA came and Raymond was asked to recite the Pledge of Allegiance before the festivities began.

Now, at age 93, Raymond sits and ponders the good old days. He still wears his Navy hat and his Honor Flight polo shirt whenever he goes anywhere that a military presence will be. Raymond has lived a good long life, endured an incredible journey and I (Rod, his son-in-law) implore anyone reading this to always remember, always honor, and always thank whenever you see a member of the military. Some, like my wife, were lucky to have dad come back home.

Stacey Walker with Lauren and Ian Stochl
The Story Doesn't End Here

In the words of Ian and Lauren Stochl

They are the themes that make the world go round. Love, life, and death are concepts humans can never escape, nor would we want to. They are the reason we exist. They are the reason we tell stories. The art of storytelling is a personal response to life. It hints at places and times in history where the story and the storyteller have been and lets us know where he or she is now. Storytelling emphasizes the human connection we all crave. Stories put us in context and connect individual bits of information to the bigger picture of who that person is, increasing our understanding of the details of his or her experience. Anyone can read a story, but when a story is told, people feel a bond between the teller and themselves. When Ian and Lauren shared their story with me and my husband, they became a part of our identity. They passed along a bit of themselves. Storytelling is an act of generosity, and even when the pages of this interview stop, there is no beginning and no ending. Ian and Lauren's story does not stop here.

For the world outside of my small suburban house, May 19th, 2018, was a day of the royal wedding, but for me, my husband, and our five year-old son, Adam, it was the day Lauren, her husband, Ian, and their son, Ian, were coming over for an afternoon of good food, beer, and playful kids. Adam woke up that Saturday ready and waiting for them to arrive and sat looking out our huge front window eager for an afternoon of play. I was looking forward to getting to know Lauren's husband and hoping our boys would get along.

I had gotten to know Lauren over the past four months because she was a student in one of my composition classes at St. Louis Community College. She stood out because she sat in front and was always at least thirty minutes early for every class. Not only was she prepared for every class, but she was wanting to learn. This always makes it easier to connect with a student; however, I am not one to shy away from someone willing to talk. And she was willing to do that. Every Tuesday and Thursday we would discuss our lives outside of class, our sons, our pet peeves, and of course our husbands. When I found out Lauren's husband was a veteran as well, I was even more connected to her. Even though we would share some similar experiences with our husbands, I soon discovered that our journeys were very different. I would soon discover that her husband, Ian, was terminally ill and living with cancer. I was compelled. I left that day from class changed. I couldn't stop thinking about Lauren and Ian and their son. I talked to my mother about them. I talked to my husband about them. I talked to anyone I knew about them. I just couldn't stop thinking about Lauren and her husband.

Because my husband, Kent, and I are so involved with veterans and helping them tell their stories, I thought interviewing her and her husband would be perfect for *Proud to Be*. I sat on it for about a month, but I spoke with her on several occasions about what we did for veterans and how we facilitated a Veteran's Writing Workshop with the help of the Missouri Humanities Council and St. Louis Central Library. It was during this time that she began working on a project I assigned in class. This was a very research-based and extensive project that required a lot of time and personal investment. I encouraged Lauren to do her project on a topic she felt passionate about, and of course her husband and his diagnosis of terminal cancer was one of her passions.

*

We stopped for a moment, and Lauren and I listened for our sons. They were playing downstairs and our conversation upstairs at the dinner table was getting loud, so checking in was the best thing at the moment. "You all right?" I asked. Adam declared they were, and so we moved on in our conversation.

*

"My son will know that if you're sick, you can still do anything," Ian declared. He was sitting across from me drinking his Guinness. In 2014, Ian was diagnosed with follicular non-Hodgkin's lymphoma in Stage IIIB, and even though he wakes up every day feeling around his neck, under his arms, and around his groin waiting for his cancer to grow again, this is not his life. He's just 33 years old and has a cancer that's only been known to affect older individuals around the age of 65. He's an expressive storyteller. His hands moved when he talked and he shifted in his chair when he was excited. I must admit most of his storytelling was exciting. He's not what one would call tall, but he is built in his arms and legs and reminds his wife, Lauren, of a "bulldog." I found myself hanging on every word that floated out of his mouth. Even my husband was clipped with anticipation. I made note of how Lauren watched her husband. Her eyes were bright and there was no escaping her devotion, understanding, and adoration.

Getting all the details of Ian's service down was a team effort. Even now, I'm looking at my original notes, Ian's handwritten timeline scratched on the backside of his VA St. Louis Healthcare System appointment reminder, and Lauren's own notes sent to me via text message. I can't help but smile at how important it is we tell this story. All of us together feel the desire to have it out there, knowing it's ready to be told.

After graduating high school in 2003, Ian needed a plan for college and joined the Army in 2004, doing his OSUT at Fort Leonard Wood, MO, 74th Delta, as a Chemical Operations Specialist. After finishing his OSUT, Ian returned home and served in the Nebraska National Guard in the 173rd Chemical Recon Unit.

Ian and Lauren's story began in 2005 when she was 18 and he was 20 and they began dating. I guess one could say that their relationship began earlier than that because Lauren was friends with his cousin, so she was familiar with this ruggedly handsome young man when she was 14. They had only been dating for about six months, when his unit received their orders to deploy in 2006. He did his training at Camp Shelby, MS, and combined with the 755th Chemical Company and left for Kuwait in September of 2006. Once he left for the Middle East, his company fell under the command of the 13th COSCOM, and when he made it to LSA Anaconda, he started running missions as a machine gunner/driver. Later in 2007, the 13th COSCOM switched to the 106th Transportation Battalion.

It was before he left and during his tour that both Ian and Lauren decided to put their romantic relationship on hold but remain close friends. This meant a lot of letters back and forth and phone calls that were interrupted by bomb blackouts and periods of not knowing what was happening and if he was okay or not. Life at home for Lauren kept going. She has on many occasions proclaimed that both her and Ian did not make the best decisions when they were younger, which I think we all can relate to. However, she is very much a nurturer, one of those individuals who thinks she can help those in need. She wants to be there for others when they "need" her, and I can identify with her because I, too, am this type of person. We are both compassionate and generous with our hearts almost to a fault. This, too, she claims is the reason she was drawn to Ian. Not only was he handsome, but he was someone who needed help, and she wanted to be there to help. And even though she knows she can't fix his cancer, that she can't eradicate the malignancies marking the inside of his body, she has in many ways fixed it already because she is here now, because she is his wife and the mother of their most glorious boy, and she is helping me tell his story.

Even now, the readers might want to help, but we just can't fix what went wrong for Ian (years of denial from the VA, years of being labelled a drug-seeker, and years of being told "it's all in your head"). But to say his journey is all bad is also not telling the full story either. The tragedy Ian faced in war is not unlike most stories from soldiers. There are the deaths of friends that served with you: those you trained with and lived with in the most life threatening situations. In these moments, you become cautious of routine and aware of superstitions and rituals, the repeated patterns of behavior that make survival in these intense situations bearable, controllable—even though you know they are not. Part of Ian's story is of course one of those tragedies. It is how he lost one of his dearest friends and comrades when he lost Sgt. Jacob Schmucker, aka "Jelly," in July 2007. They were a team. There was nothing they couldn't accomplish together. The two of them went on mission after mission and both survived when others didn't. They were each other's "lucky charm." They felt safe knowing the other one was there. However, on July 22nd, 2007 this would change, when Jelly went on a mission without Ian because he had been delayed on a mission in Baghdad. Even though Ian knows the reason for Jelly's death had nothing to do with him not being there, the feelings of guilt still ring true. Upon Ian's return to find out that Jelly was killed, the truth of it seemed like a cruel lie. It was several months later that Ian was asked to lead his last mission north to checkpoint 59.5 Alpha before heading home, back to Nebraska.

In his own words…

"We were only about two weeks from going home. I got a knock at the door and it was a runner from the Tactical Operations Center (TOC). They told me that the First Sergeant and the Company Commander would like to speak with me. At first I was worried I was going to get an Article 15, so as I took the bus down to the motor pool where the TOC was located, I tried to remember what I had done. I also tried to think of a reasonable explanation for what I might have done.

"This was not the case at all. I knocked at the door and requested permission to enter. I was greeted by not only our CO, but also the CO for the company that was replacing us. Knowing our company would not air our dirty laundry in front of anyone else, I immediately realized that I was not in trouble. I was asked if I would volunteer for one more trip outside the wire. However, this time it would be as a driver for Gun 6, our new Roving Security Vehicle. Even though this was not an order and I didn't have to go, I was reminded that the company that we replaced went on a few more missions to show us the ropes, and a lot of the tactics they had taught us kept us alive many times over. Honestly, the first thing they had taught us was to forget the shit you learned stateside because that would get you killed. In the end, they allowed me a day to think it over, but I knew that the information that I had acquired over the last year wouldn't do me any good back home and would probably keep some of these newbies alive. So, I accepted the mission.

"I was given the keys to a brand new 1151 Humvee with only fifty miles on the odometer. All the missions prior we had run M998s with whatever armor we could bolt on until the actual armor kits came out, which sapped all the engine's power. The new M1151s were redesigned from the ground up to hold the massive amount of armor that was attached to it, with upgrades like a turbocharged engine, new suspension, and most importantly AC. The mission itself was to be a short run up to Warhorse and back, north of LSA Anaconda on MSR Tampa. Unfortunately for us, what used to be a short and relatively safe trip had evolved into a much more dangerous ride. The fighting had shifted from the south to the north, where 59.5 Alpha used to be the most dangerous checkpoint we rolled through, but it seemed to have died down.

"It started as every other mission did, we had a quick formation and safety gear check, ran PMCS on the vehicles and weapons, and ran by the test fire pit to make sure everything shot like it was supposed to. We went out the north gate and proceeded north on Tampa. The sun was setting over the desert as it almost always did on our missions. We even had the iPod hooked into the comms system so our vehicle could hear music through our headsets. We were off. Everything seemed as normal as could be for a war-torn shithole until we started approaching the landmark we called the "Stairway to Heaven." It was a strange monument that the Iraqis had built stretching over their highway. It had two arches topped with golden domes, and in the middle there was a spiral staircase leading to a small room, which overlooked the highway in both directions. Just to the south was an Iraqi Army checkpoint with standard Jersey barrier serpentines leading both north and south with the actual Army post in the middle of the two.

"We sent Gun 1 through the checkpoint while Gun 2 held back the TCN drivers. As soon as Gun 1 made it to the other side, our radio came to life, 'Gun 6, this is Gun 1, over.'

"'Gun 1, this is Gun 6, go ahead, over.'

"'Roger, that. We saw a pile of shit on the back side of the Jersey barrier just past the hut, over.'

"'Gun 1, could you define 'pile of shit,' over'

"'It looked like a bunch of big plastic soda bottles filled with liquid, over.'

"'Roger that, we'll check it out, over.'

"'Gun 2, you catch all that? Over.'

"'Roger, we will hold the TCNs until all clear, over.' So, we cautiously approached the Jersey barrier in question, just as Tupac's 'Picture Me Rollin' came onto the iPod. Not being used to being a driver, I was leaning over the radio, trying to get a glimpse of what the hell we were looking at. My gunner Sgt. Noyes was leaning over the turret armor with his Surefire to light up the pile. As soon as his light hit the pile, our worst fears materialized. It was a Russian-made 155mm artillery round wired up with a bunch of plastic bottles three-quarters full of yellow liquid, and our front tire was damn near on top of that son of a bitch.

"Training kicked in, and I swept the gunner's legs out from under him just as the device detonated. If there would have been a second of hesitation, I would have only been pulling half of my gunner in. It was so close that the cuffs of his blouse were singed. There was a white searing light, and then blackness. I don't know how long I was out for, but I started thinking again very shortly after. Everything was still black and I thought about how I knew it: no heaven, no hell, just blackness. This thought train was just the last bit of electrical energy stored in my brain and would soon subside into the black abyss of death. But it didn't, and I kept thinking. So, my first cognitive thought was I am just blind and deaf, not so bad, I am still alive. Then, I thought about my hands. Oh, I have hands, good. Check my legs. Oh, got my legs. Check my stuff. Got my stuff. So, I thought I was blind and deaf, but I had all my parts, not so bad.

"Then my sight and hearing started to fade back in. The first thing I saw in my smoke-filled Humvee was Sgt. Noyes rolling around on the ground screaming that he was hit and dying. I start to pat him down looking for blood. There isn't any, and it takes some convincing to show him that. Next, I try the radio, which is totally screwed. Then as if being told by a higher power, I try the wheel and gas, and although half my engine was vaporized and there were no tires left on the rims, I smashed that gas pedal, grabbed the wheel, and looked through the one inch of my windshield that was not blackened, and much to my surprise the Humvee lurched forward in very awkward jerking motions.

"I was just barely able to see the brake lights of Gun 1, so I pulled Gun 6 alongside of her, so we would have cover from Gun 3 and Gun

9. Half of the front of my Humvee was gone, like it had never existed in the first place. We hooked up Gun 6 to the wrecker, and I hopped into Gun 1 and Charlie Miked to Warhorse. We finished our mission, dropped our load, and I rode Gun 1 back to Anaconda. My thinking was I couldn't get smoked twice in one night. Once home we did the IED evaluation from the hospital, and we told them that we were fine. That was not the case—I was having some serious headaches, memory problems, trouble sleeping, fatigue, and all kinds of other shit, but we feared that if we were injured that we didn't get to go home, but would have to go to Walter Reed, and that was a fate worse than death. So, we lied."

*

Even though that lie allowed Ian to come home, he would, years later in 2011 while seeing a doctor at Jefferson Barracks in St. Louis, find out that he was affected with Traumatic Brain Injury from the 155mm IED bomb on November 2007—PTSD, the damage to his lungs from being hit in a gas attack in August 2007, and damage to his back from the weight and wear of his body armor. But this was only the beginning to Ian's body breakdown.

Once Ian returned to Omaha in December of 2007 after serving his time in the military and in the Middle East, he spent the next year getting into too much trouble drinking and doing drugs trying to self-medicate because he couldn't get help from the VA. He was led to believe that in being a veteran there would be tremendous opportunities and that "everyone wanted the vets!" This was not the case, and desperate for help he heard that St. Louis Jefferson Barracks had a new Polytrauma Unit. Since he had family (his grandparents) in St. Louis, and knew Lauren was also there, he came to St. Louis in 2009 where he began pursuing Lauren again.

And like all great stories, she declined his advances. She would tell me she knew that if she let him in that that would be it. They would never look back, but she wasn't ready, and she knew he wasn't either. However, I know from past experience that sometimes we are just drawn to the fire. Besides, they never really stopped being friends. She knew she was different and she knew he was different since coming back. He was more cautious, more introverted, and in many ways less emotional. Despite all of this, however, he was also just a little bit lost. He needed help, and she knew she could do that. That's what she was good at, and so in late 2009 he finally, "locked Lauren down." Through all the chasing, the war, the wrong choices, the moving, the pain, and what was about to come, he had her and she had him.

"Imagine at 30 years old a person should be enjoying the highlights of their life, working hard, and planning for their future. Imagine a young father in the peak of his life. At the age of eighteen he went to fight for our country at a time of war, during Operation Iraqi Freedom (OIF) as a proud member of the United States Army. Unlike some of his comrades, he was lucky enough to come home. My husband completed his seven years of service to his country. He went back to school, we got married, and had a son. There was nothing but promise in his future. Right after he returned from his mission in Iraq he was bombarded with medical issues. He was unable to sleep, had terrible night sweats, chronic headaches, and extreme back pain, not to mention the fact that he had psoriatic arthritis so bad that at times he had to use a cane to walk. For years after his return, he went to the doctors and discussed all of the ailments that were plaguing his body. His doctor, through the VA, overlooked all of these problems and told him there was nothing wrong with him. They told him that it was psychological and it was 'all in his head.' He spent many years struggling and in pain. He was told for so long that nothing was wrong that he started contemplating if it really was 'all in his head.' Years later, and after many visits to the VA, he was finally sent to see a specialist for imaging. Before leaving that appointment, he found out that he had cancer. Unfortunately, by this point the tumors had metastasized throughout his entire body and into his bloodstream. Radiation was impossible. The fact that it took so long to treat him makes his cancer incurable."

*

But before the discovery of Ian's cancer, the story of Ian and Lauren was about to change on November 30th, 2011. Their son, Ian, was born, and with it a new future. Even though Ian was still very much sick, he worked hard in construction, and in 2012 he enrolled and started school at St. Louis University. This was the main reason he joined the military back in 2004, and despite being seriously sick and not knowing what was wrong, he was a father, he worked full-time, and went to school full-time.

After years of being told it's all in your head, after years of not sleeping, after years of physical pain, after years of swollen tumors around his neck, under his arms, and around his groin, in the fall semester of 2014, Ian was finally diagnosed with follicular non-Hodgkin's lymphoma in Stage IIIB.

After eight years of being dismissed, Ian and Lauren finally knew. They had planned to get married within the year, but hearing this news, they moved their wedding up, and after his first chemo treatment on November 14th, they were married on November 21st. Ian began his first six months of intense chemo treatment. He was injected with two different drugs once a month, and later it switched to one drug every month. In the spring of 2015, he decided to take six months off from school to continue his intensive chemo treatment. Again, the treatment consisted of a round of chemo every four weeks, then every eight weeks, and finally, every three months, and in the fall of 2015, Ian was back in school and still getting chemo treatments. In the beginning, the treatments were on a Monday, and Ian would have to go to his classes the rest of the week without any recovery time. However, his treatment changed to Friday, which would give him some reprieve before heading back to school. There were times when he would leave his chemo treatment and head straight back to class afterward. His final chemo treatment was on August 20th, 2017.

Ian graduated from SLU just three months earlier in that spring of 2017, and started a new job as an Energetic Nanoparticle Chemist. His son, Ian, just finished his first year of school and is in the gifted program. Lauren is now going back to school to get her degree because she knows the reality is that Ian won't be here forever. She fears being a single mother and not having the ability to raise and provide for her son. I would too. The outlook for Ian is at most 15 years. Their plan is five years at a time. They know the cancer will never be gone. It sits there under his skin and is going to grow again. They know when it does, chemo might not work, but maybe there will be something else that will. They will not give up.

I asked Lauren what she wanted the readers of their story to remember most, and she wrote some thoughts down one night while in bed: "Because the clock on Ian's life is set, we appreciate each day. Having cancer has made him work harder at succeeding in all aspects of his life, and he is determined to set the example for our son." In the end, they are making a life worth living for and a life worth sharing. The readers of their experience are clear evidence that this is not the end but only the next day, the next hour, the next minute on this family's precious clock.

Kill or Be Killed

"It was kill or be killed." This was the response to the second question I ever asked my dad about being in the war. We were on our way to Florida, driving from Connecticut, with the trailer in tow. My mom had run into the gas station mini-mart and my dad and I were both still trying to wake up from our mid-trip nap. I was about 12, and we had never talked about the war. It was Mom that always put the kibosh on the discussion. But now my dad and I were left alone and to our own devices. There was a scar on his leg that I had not noticed before, and I thought it may have possibly been the result of what infamously became my mom's tale of how while building our house, he almost cut his leg off with the circular saw. My dad, Fred, was easy-going and we had a great relationship and rapport.

I flipped up with my finger the hem of his denim shorts and asked, "Oh, is that from the circular saw incident when you almost cut off your leg?"

"Noooo." His voice and his crumpled expression dripped with disdain for the memory. He switched tone, "You've never seen this before?"

"No. When do you really ever wear shorts?" I joked.

"I got this in the war."

Now he had piqued my curiosity, but my mom's voice was in my head, saying, "Don't talk about the war. He has bad memories of the war." Youth and naivety pushed me to ask. Just as the question was leaving my lips, my mom returned to the car. She was standing outside with a hand on the open door. You could hear a slight gasp as she returned in time to hear the latter part of the inquiry.

My dad asked my mom, "Why do you think I don't want to talk about the war? I want to talk about the war."

All parties discovered that day that my mom, in trying to protect my dad, was telling all of us not to mention the war to him; it was that day that we ended the practice and he talked openly about his experiences. The first annotation involved the three scars on his upper leg.

Fred was a member of the 280th Combat Engineers Battalion, Company A, that was a part of the larger 3rd Army under the command of General George S. Patton during the Second World War. He, his buddy Jimmy, and another soldier were on patrol, walking through a field in France when an 88 exploded near them. Under fire, the third gentleman left for medical care while Jimmy and my dad continued on. One of their commanding officers (from HQ), not far from their position, unfortunately lost his life when the jeep he was driving took a direct hit. I asked my dad why he didn't leave to find a medic. He said, "We had a job to do. It wasn't that bad and neither were Jimmy's wounds, so we kept going."

Not too long before, I had seen a news story about a WWII veteran who completed all of the documents and was awarded his Purple Heart. I asked my dad if he wanted me to do this for him; he was wounded in battle and he deserved to receive the medal. He just shook his head and told me, "We were just doing our job. That's what we were there to do. We didn't have time to go for medical treatment." There was no glory associated with it. It was their job; they did it well and they did it to survive. Although it didn't matter to him, I have always felt some guilt about not going through the process for him.

In honor of the 280th's 50th anniversary, a good number of the remaining members held their first reunion at Camp McCoy, now Fort McCoy, in Wisconsin. It was here the men of the 280th completed boot camp before shipping off to Europe. Due to work commitments, I wasn't able to accompany my dad. Two years later the second reunion was held in St. Louis, Missouri.

I had the opportunity to attend the third of the battalion's reunions; this particular one happened to be in Dallas, Texas. Arriving with my dad, I was quickly acquainted with all the men whose names were ingrained in my head, and I was able to pair the faces with personalities. During the planning of this trip, I knew that it was going to be informative, but I did not truly realize how much I still did not know, nor understand, about my dad's time in the service.

In the conference room of our hotel, we were seated at large round tables. Fred was the proud father and as he was introducing me to all of his buddies the men would come up and shake my hand and speak with me a bit. But, they kept saying something strange, at least to me it was odd. Each one would say, "Your dad is a hero." Finally after enough times, and holding back my emotion, I said, "Yes, he is. But, you all are heroes." When we were all seated, we happened to be at the same table with two men, Jimmy and Ken, who were not only very good friends of my dad but who had served side by side with him during the war. Both had heard my response and were surprised that I didn't understand why they were all saying this to me.

Among the lighter, funnier tales, and, to explain why they thought my dad was a hero, the boys decided to tell me the story of how they, Jimmy, Ken, and my dad, were on patrol one night. My dad, as he was accustomed to being, didn't want them to divulge the details. He didn't want to be in the spotlight. But, they persevered. It was a dark, moonless night. As they made their way down the dirt road they could hear the sound of voices coming from the not-so-distant forest. Those voices happened to be speaking German and they belonged to an enemy patrol unit.

Running along the road was a stone wall, knee high. As they sat on that wall, they could hear the voices getting ever closer. The idea was to jump behind the wall and hide just out of sight. Presuming it was only a meter high on the opposite side, they could stay hidden while being able to observe the impending situation. Well, because of the darkness they couldn't see and all three jumped and fell about 10 meters. One of them hurt his ankle, the other his back. Not being able to stay there with the enemy quickly approaching, my dad made his way back up to the road.

Needless to say, the three of them had to think fast and had to act even faster. The blackened night may have been a bad omen at the beginning of this adventure, but in the end, it turned out to be quite the extraordinary accomplice. At this point, they said to me, your dad, that night, captured the entire German patrol. I cannot exactly recall, but I think all men seated with us chimed in together, "And, that's why your father is a hero!"

It still is difficult to match the words to my facial expression. I just looked in disbelief, in awe, my mouth agape, not because I couldn't believe that my dad was capable of this, but that he never told me the story. Fred sat there with his arms folded against his stomach and chuckled. My mouth still opened as my eyes darted between him and everyone else. I finally said to him, "Oh my God! You captured all of those Germans? You're a venerable Sergeant York!"

During my life, I had been told of the many adventures that my dad and the rest of the company had, but hearing them there, with the actual parties that were involved, made it all the more real. As combat engineers, the 280th was a battalion that was responsible for building bridges and roads and cleaning those roads (especially of snow), and much of this work was done under fire since the Engineers had to go in to clear the way for the infantry and tank divisions.

After the battle, the Engineers went behind those same groups to re-collect the U.S. Army materials, but also to make sure the enemy soldiers were dead. In this process, many personal and historical items were collected. My dad, for instance, brought home an SS Cross medallion, a Walther P38 pistol, and a French-German mini dictionary, along with several other cherished pieces of memorabilia. It was during this conversation that I asked my dad what it was like to kill another human being. He was a hunter and I followed in his footsteps; the sentiment in hunting an animal used for food is far different than killing a person. In no uncertain terms, he explained, "He was the enemy. You either killed him or he killed you."

Life was difficult and these boys were a long way from home. A good many enlisted, but the majority were drafted, barely eighteen years old, and fighting a war on foreign soil. There was bound to be brevity, to say the least, during their time in the service. My dad, Jimmy, and Ken, as I mentioned, worked closely together, as the three were assigned to a 2½-ton truck. As civilians, we may take a great deal for granted and don't think of the hardships of being in the field without any of the conveniences of home. To make their lives a tad bit easier, the three of them devised an apparatus.

I remember my dad asking, "You've seen the trenches on the side of the roads?"

Not knowing where this was going, I answered, "Yeah, they were dug so you could jump in them if you were under fire, no?"

With his famous Fred chuckle, he said, "well, yeah, but we used them for something else." Of course, I had to ask because I was curious and he had my attention, and at the same time, I was afraid to know the answer. Going to the bathroom was an uncomfortable nightmare when you had to just squat on the road, in a field, or beneath a tree. To create something more convenient, they took a sheet of plywood, wide enough to sit over a roadside trench. In that piece of wood, they made three circular cutouts. Kept in the back of the truck, they could easily pull it out and put it over a trench when nature called and it made the entire process a bit more comfortable.

Another great tale that my Dad told me, and I actually asked Jimmy about it to make sure it was true, happened one night between the two of them. With no patrol to go on and having nothing much to do, what do 18-year-olds in the field want to do? No, not women. These guys wanted to drink. My Dad and Jimmy had the bright idea to head into town to get some liquor and to bring it back for all the guys to imbibe.

The evening started with Jimmy and Freddy stealing a motorcycle from the "Air Force guys". I did inquire as to who these "guys" were or from what part of the AF they hailed, but neither could remember any details. With a motorcycle now in their possession, the two of them headed into town, which was in France. While there, they both loaded up with bottles of cognac, and I am sure they indulged a bit, too, before heading back to camp.

Time passed and they started to make their way down the dirt road full of holes and uprooted rocks. Something else they weren't expecting, since they were returning later than they had thought, was that there was no moon. It was completely black. And, they couldn't use the lamp on the bike because they were under a blackout order. As they sped along the rutty path of a road, the bike's front tire hit a hole straight on. My dad was driving and Jimmy was behind, and both went head over heels, flying over the front of the bike. My dad showed me a scar on his chin and informed me that it had happened when his face slammed into the handlebars. He also joked that he, Jimmy, and the bike were a bit banged up but none the worse for wear. And more importantly, "Not a drop of cognac was lost. We didn't even crack a bottle!"

The 280th Combat Engineers saw its share of fire and was an integral part of several major battles. The most famous, and one of the most important, was the Battle of the Bulge. My dad would always say that any film that tried to recount and reconstruct the battle never was able to accurately show just how much snow there really was. The Engineers were trying to clear roads as tanks and other heavy equipment were slipping and sliding, losing absolutely all traction. Ken's wife told me that he ended up getting frostbite on his feet during that battle in the Ardennes. When they were discussing where they wanted to settle down after they were married, she told me that he had only one request, he wanted to live where it was warm. He had no qualms about taking a vacation where there was snow, but he, in no way, wanted to live in a place where they had a real winter and where it regularly snowed.

I thank my dad, and all of the members of the 280th Combat Engineer Battalion, for being able to convey all these great stories and to pass along the memories to me so I can then tell them again, to keep the history of this era alive. Fred passed away in 2003. Not only did I lose a loved one, but we all lost a part of history, one that needs to keep being retold. This world, especially the younger generations, needs to be reminded of the true history that lies within each and every person of the World War II era. The men and women who served will soon be no more, and it is our responsibility to continue to tell their stories.

I lovingly dedicate this piece to all of the men and women who have served, who are serving, and who have laid down their lives to protect their citizens and the hard-fought freedoms that we possess. Wherever he may be, I again thank the hero in my life, Frederick M. Ferguson, 280th Combat Engineer Battalion, Company A.

Jay Harden

How I Survived the Bulge and Other Stories:
An Oral Interview with Uncle Sam

All veterans have a close relation named Uncle Sam; we work for him and what he represents. A Vietnam vet friend of mine also has an admired uncle, a real Uncle Sam. Samuel E. Losh, Jr. is 93 and a veteran of World War II. He recently visited Washington, DC, with his grandson, courtesy of Honor Flight. Honor Flight is a non-profit organization that transports veterans at no personal cost, to see their war memorials in Washington. That experience breathed fresh vigor in Sam and ignited his memories, now all the more important to record since the Military Personnel Records Center fire of July 1973 in St. Louis destroyed his service records.

After his trip, Sam finally decided to preserve his own Army stories for family and friends in an oral interview with the author on October 16, 2017. These are Sam's memories that he wanted to record with me.

Sam remains bright-eyed and bushy-tailed with a persisting sparkle for life. In particular, Sam remembers why he didn't die in the Battle of the Bulge.

Sam Losh at 93 in 2017

Sam worked as a welder's helper making bomb racks for the military and could have gotten a deferment, but his brother and buddies had already enlisted and he came from an extended St. Louis family of several Navy men. Sam tried to join as a sailor, but was disqualified for color blindness. His stepfather and brother were in the Army, so

Sam went to Jefferson Barracks and volunteered at age 19 to wear green instead of blue. That was in December of 1942.

Sam ended up on a train bound for Camp Callan, California, 15 miles north of San Diego, and in 13 weeks of basic training in anti-aircraft artillery.

Camp Callan, California

Sam qualified as a small-arms expert on the World War I Springfield 1903 .30-06 bolt-action rifle with grenade launcher attached and ended up assigned to the .50 caliber machine gun that supported and defended the cannon crew. He also trained in the Mojave Desert on the old 37mm autocannon left over from WWI that jammed a lot. Sam's job was to lie under the cannon and catch the very hot unfired rounds and pass then to another man to put in a misfire pit. Lucky for Sam they started training on the 40mm Bofors anti-aircraft automatic cannon intended to protect the nearby field artillery from air attacks.

Bofors 40mm automatic cannon

He graduated from basic training on January 9, 1943 as a member of the 462nd Anti-Aircraft Artillery Automatic Weapons Battalion (462nd AAA AW Battalion) and started his journey to the European Theater of Operations (ETO).

Sam Losh at 17

After graduation, Sam was sent to Camp Haan in Riverside, California, as a replacement gunner in a new unit just starting their basic training. He was told he had to repeat all of basic training, but Sam angrily refused and was disciplined with a new occupation in Kitchen Police (KP).

In our conversation, I asked him how he kept his sense of humor during war. "I don't know. I was just a kid and everything was funny to me. Even the obstacle courses were fun."

Camp Callan Obstacle Course

Sam, to this day, still has that positive perspective of an immortal 19-year-old. He mysteriously told me early in the interview, "If I had it to do over again, I'd prefer KP."

"Why?" I asked.

"Well, as I will tell you, KP kinda saved my life."

In April, Sam headed for the U.S. Army Amphibious Training Center at Camp Gordon Johnston in Carrabelle, Florida, where he learned to jump from a burning ship hanging from a cargo net on the side while the landing boat bobbed with the waves up and down and away from him. He also learned to swim through and under flaming oil on water.

In May, they pitched pup tents in the mud of Gallatin, Tennessee, for simulated war games with the 79th Infantry Division. The maneuvers seemed like a furlough to Sam and the hardened men of Battery A.

In June, they were trucked to Camp Stewart, Georgia, for more weapons practice and training in invasion disembarkation from a Higgins Boat. Once, Sam jumped from the boat and sank in a big hole while the following guys jumped into the water on top of him. Sam nearly drowned and had to abandon his gear so he could finally surface. The officer above asked, "Where is your rifle and pack?" Sam had to dive down again and retrieve it all.

During July, that battalion went to Camp Pickett, Virginia, and the luxury of barracks, bed, and showers as the truck drivers practiced black-out moves. They also spent a week on the Atlantic coast at Camp Bradford, Virginia learning to use the landing craft that later cracked the beaches of Normandy.

While home on a three-day pass, Sam called his 17-year-old girlfriend, Laura, and said, "Pick out your best dress and meet me in the morning. We are getting married." Laura's stepfather was in politics and got the three-day waiting period waived. Sam and Laura drove around until they found a Lutheran minister working in his church garden and he agreed to marry them. Laura's mother and Sam's sister and brother-in-law were witnesses. At the end of the pass, Sam reported for duty. They did not see each other again for almost two years. Laura is still the loyal love of his life after 74 years of marriage.

His point of overseas departure in the States was Camp Myles Standish south of Boston, Massachusetts. There the 462nd joined up with the 17th Field Artillery Brigade. Sam learned the art of freezing in his summer fatigues, a ruse to mislead spies into thinking they were headed for the Pacific. There, while peeling potatoes during KP, Sam met an old mess sergeant who gave him his life-saving advice. He also mastered the art of "hurry up and wait" and confessed that he stole officer steaks on occasion for private consumption.

After 11 months in the U.S., Sam left Boston for the ETO on November 6, 1943, in a converted freighter, skipping wave over wave across the Atlantic and practicing personal combat with another unseen enemy, seasickness. Sam stayed well most of the trip until three days before port in Great Britain where they finally received warmer clothing.

Laura and Sam Losh in 1942

When allowed, the G.I.s visited the local pub in Pembroke, Wales, to relax and drink ale. The girls in this poor lumber town naturally sought out G.I. money for free drinks. (Like the other privates, Sam's base pay was $50 a month; he kept $5 and divided the remainder between Laura and his mother.) G.I. generosity did not sit well with the local testosterone of miners and lumberjacks. "We got beat up regular," Sam said. So one morning at 2 a.m., the captain came to the barracks and ordered, "Fatigues, boots, helmet liners, let's go." He led his men to administer some physical attitude adjustment to those young men. From then on, Sam recalled, they had no trouble spending money in town.

The 462nd arrived on the east coast of England just before D-Day, ready for battle.

Seven days later, June 13, 1944, Sam and Battery A landed at Omaha Beach, Normandy, in an LCT (Landing Craft, Tank) with his .50 caliber machine gun, their truck, and 40mm cannon in tow. Their first sight in Europe was Allied bodies stacked on the beach beside unexploded mines and other debris of war. That memory still brings tears to Sam's eyes.

Sam recalled, "I had some crazy experiences there. Thank the good Lord I made it."

For example, Sam said, "My first combat was with a spider. When I landed in France I started to go deaf. Then the medic found a dead spider in my ear."

Sam and Battery A lived in the field, constantly on the move. "Once it rained for 30 straight days. I slept in a slit trench one night and woke up with only my head on my helmet above water."

He remembered one humorous pleasure. "In Belgium, 250 of our men walked naked under a long pipe of hot water with holes in it. At the end we had new uniforms waiting. The local farm women watched and laughed."

The 462nd scored their first combat kill four days after landing at Normandy. After six weeks, Battery A had shot down three planes. The Battery had a crew of five: Sam, the defending machine gunner; the cannon gunner; the shell loader; the elevation man; and the azimuth man. Soon enough, Sam helped his crew shoot down a Nazi fighter, probably a Messerschmitt.

"One of the worst experiences I had was scanning the sky and watching five B-17s get blown out of the air. Had to turn my back. Couldn't watch it. No parachutes."

"My other job was a communications lineman and wire runner. I ran a 5-mile reel of wire from our gun to the Command Post. At 3 a.m. one morning, I took a shortcut across a field running my wire. I remained overnight at the CP, then returned the next day to my gun via that field. The sign said: 'DANGER MINE FIELD!' I sat down against the fence and threw up. From then on, I never cut corners and never ran wire alone.

When they came up with battery-operated handheld phones, I said, 'Thank you, Lord, no more wire, no more radios. Thank you, Lord, for that.'"

Sam recalled another adventure: "Once we had a rest camp near a brewery town. So, we went looking for beer and found a keg, then tested it with our canteen cups. Neither one of us was able to speak; it took my breath away. Later I found it was Calvados apple brandy of 150 proof (75% alcohol) like bootleggers make."

I asked, "Make any mistakes over there?" Sam chuckled. "I was a scavenger: I found some jars of canned fruit in an old farmhouse cellar and ate a jar of apples; and spent five days in the toilet. Also, got my eyes windburnrd while driving and watching, without goggles, for enemy planes to strafe us. I was blind for several days until my eyes healed."

"At a rest camp in Belgium, a farmer went digging in his apple orchard and brought me a .32 pistol, rusty and corroded, wrapped in a rotting rag. It still worked. I was carrying it in my backpack when I got hit."

In October of 1944, Sam and Battery A were ordered to a quiet sector in the Ardennes forest where, unknown to them, the Battle of the Bulge would soon begin.

Battery A had the mission to protect the 15th Field Artillery Battalion as they advanced behind the 2nd Infantry Division of the 1st Army, and Patton's 3rd Army, against the German 7th Army and 6th Panzer Army.

2nd Infantry Division Patch

During the Battle of the Bulge, the encircling Germans had trapped many American soldiers as they tried to get back to Allied lines. On the second day of the battle, December 18, 1944, at 2:30 a.m., Sam told me, "I got blowed out of a foxhole with a hand grenade."

Battery A was sleeping in an abandoned farmhouse near Saint Vith, Belgium. At 2 a.m., Sam went forward to relieve a guard on duty in a foxhole, watching the snowy woods for approaching enemies. He heard the sound of rocks coming down the road and shouted out the challenge of the day, but got no response. Then a German potato masher grenade sailed past him and exploded, lifting him out of his foxhole.

The other G.I. guard ran up and jumped on top of Sam, whispering, "Sam, are you hurt?"

Sam answered, "I am now!"

Then the officers came running and asked Sam, "What color were their uniforms and what kind of guns were they carrying?" Sam answered, "Sir, it is 2:30 in the morning, black as the ace of spades, so how am I supposed to see the color of their uniforms?"

Other G.I.s lifted Sam out of a trench, then placed him face down on a stretcher across the hood of a Jeep, and drove him to a first aid station in a nearby house for stabilization by medics, then to the field hospital.

"When I was wounded, I never lost consciousness. I remember the medics counting out loud how many holes I had in my back, butt, and legs. Their grenades were designed to cripple you, to take you out of action. The Germans believed that would also take support people out of the fight." He added, "The German machine guns were bolted down knee high to cripple us." The Germans believed that one wounded man would reduce the total number of fighting Americans more than one dead G.I.

Sam remembers, after sixteen months of combat duty, "I like to think I saved my men sleeping in that house at the end of the road. I've been carrying metal ever since." He smiled and said, "I have a lot of fun with the shrapnel in my back and butt when I get X-rayed."

Then Sam told me, "If I did it again, I'd prefer KP." Why such a curious answer, I wondered?

Sam said, "That old mess sergeant took care of me and gave me some good advice. He taught me one thing, 'Son, when you challenge someone on guard duty, turn your head, never challenge straight.' That night I challenged the German, I turned my head and that's why the grenade went behind me. I turned my head to throw my voice away from me and I guess that's what really saved me that night."

Sam was evacuated to a hospital in France. "Nine months on my belly; no feeling below the waist. I was strapped to a bed and they cleaned my wounds every day. Then I went to England."

"What was your great wish over there?" I asked.

"My great wish in combat was to get it over with and go home. I saw the Statue of Liberty through a porthole. Then an ambulance took me to Fort Custer in Battle Creek, Michigan, where I was honorably discharged as a Private First Class."

"After ten months in hospitals, I was able to walk again, and then was discharged in September, 1945, with a new uniform, $300, and a ticket home. I took a train south to Union Station, then a streetcar to my house in St. Louis, ditching my leg brace along the way, and walked into our house to surprise Laura. Coming home was fantastic."

"Sam," I asked, "What did you value about military service?"

"Discipline, respect for authority and other people. No one is different. It made me a better person and husband. I learned humility, comradeship, how to get along with people different from you, and to keep your eyes and ears open. The hardest thing I had to learn was taking orders."

"What do you want remembered about you and your service?"

"Not much. I did what I had to do: did my job, came back."

When we concluded our interview, Sam still had that look, that sparkle of freedom given and brought home from France and Belgium, an inner energy far surpassing the collection of German steel shrapnel in his body.

As I was heading home after the interview, I couldn't help but reflect with awe at what Sam and the other 19-year-olds did for me and my future in 1944 when I was one year old. Thanks to them, I got to have a free boyhood and grown-up life. This interview was my secret privilege to pass on Sam's story to you and add a little polish to our national tradition of military service that ensures the freedom of future Americans. Across a generation, as generations before us, we are brother warriors who would do it all again if need be. To others who wonder why we Americans do what we do for them and for the world, that is the best explanation I know. I think my friend's Uncle Sam Losh would agree.

Fiction

Cynthia Teramae

Gurkha!!

First Lieutenant Roxanne Fuqua left the villa late for lunch, missing the bus, and so she decided to walk the one-half mile down the dusty road to the makeshift bus stop alone, breaking rule number one—never walk alone. It was hotter than usual that day, maybe 125 degrees by noon. People who say it doesn't feel hot in the desert have never experienced Bagdad in the summer.

Sweat trickled down her head beneath her heavy helmet. One small bead glistened on the tip of her nose before she wiped it away with the cotton handkerchief she always carried in her uniform pocket, a distant memory of her southern upbringing.

She had been in Iraq for only three weeks, and the oppressive heat and Bagdad sun had already ruined her sensitive skin. Each night when she brushed her teeth with bottled water, she counted a new freckle. The handkerchief in her hand was smeared with dirt, her nails bitten red and swollen. An ugly habit. Her curly red hair was a constant damp and tangled braided mess.

She ambled along the road. Her body, heavy with the more than twenty-five extra pounds of required equipment dragging her down. She hunched forward, while the heavy ballistic body armor hung on her slender frame. A high-tech gas mask swung on her hip. Its awkward cord ran around her groin and attached at her belt. She shifted her backpack and leaned back to distribute the weight.

Her helmet always sat a little askew and dug into her head leaving a bright red line around her forehead just above the eyebrows when she took it off. A constant reminder that she was in the Army. The usual mid-afternoon headache throbbed.

She had no idea how she got there in Bagdad, and admonished herself ever since she realized that the tradeoff for a free four-year degree was a reciprocal tour of duty in the Army. Who knew there would be a 9/11? And a follow-on invasion of a country she wasn't even sure she could locate on a map. "There's never a free lunch, Sugar," her mum would say. And now she thought her mum was wise, but that new knowledge wouldn't help her now.

She surveyed the meandering water running through the gated neighborhood. It was one of the many enclaves making up the larger city of Bagdad and was located inside the Green Zone, the heavily fortified area that had been apportioned off and secured by the U.S. military soon after the invasion. They called the neighborhood "The Canals" because of the water separating the villas. The villas were huge rambling homes separated by low rock walls and once -ush gardens. Palm trees swayed among the homes. Red, yellow and pink bougainvillea bloomed everywhere. The colors collided with the now browning lawns. These were the homes of Iraq's elite, who had all fled the country now, leaving behind their grand homes. It was a true oasis in this otherwise impoverished country.

Her unit had taken over one of the villas and turned it into a headquarters for their mission. It came with a storied past—it was a gift from Saddam Hussein to the president of Iran, although they were told that the president had slept there only one night. This bit of folklore made the villa seem even more outrageous for its unused sumptuousness. The irony of a well-cared for, and still exquisitely maintained villa inside a gated community where just outside its gates, the desperate citizens of the city spent their days standing in line in the smothering heat, waiting to buy a loaf of bread or some small amount of vegetable to feed their families, was not lost on her.

Before the invasion, there was food on their tables, their families had air conditioning, at least some of the time, and their husbands' hands had not been forever soiled from having to take the only jobs available—like cleaning a toilet where a foreign woman had sat.

The villa that Roxanne's unit had taken over as their headquarters was still lavishly furnished with antiques and expensive, heavy silk brocade drapery that hung in the windows. Her unit had moved the bedroom furniture out to make room for their desks, but they kept a lot of the other furniture like the dining room table, which was big enough to hold a party of twenty. They used that room for conferences, setting up all sorts of video equipment that allowed them to talk to their headquarters back home on a daily basis. A huge industrial generator was shipped to Iraq and placed in the villa's backyard. The generator would ensure around-the-clock air conditioning and electric lighting. Such a luxury was unthinkable outside the walls of the gated community.

In the hallway was an intricate inlaid credenza that Roxanne just had to run her hands over every time she walked by. A small comfort. This piece was smooth to the touch and warm, as if it were holding the heat of the day or secrets from the past. Touching the credenza made her recall her grandmother's house, which had a very similar piece. The touch brought back memories of gumbo boiling softly on the stove top,

154

and the fresh smell of her grandmother's clean sheets billowing in the backyard.

She had arrived with a small contingent of U.S. military and a handful of civilian contractors. It was an accident that she was there, replacing someone who broke a leg just days before deployment.

Too tired to be in a hurry, Roxanne slowed to observe a ladybug nestled within a bougainvillea bloom, its tiny wings spread wide, ready for flight. She caught it in her cupped hands and made a wish, then blew through her hands to set it free. She was enjoying this small sliver of time to pretend that she was anywhere but there.

Some of the villas reminded her of the large mansions scattered along the river banks back home. As she walked, she tried hard to peer into the windows, and wondered what had gone on inside these still-majestic buildings. She stopped and closed her eyes. For a moment, she saw herself standing along her hometown riverfront in New Orleans, transported away from this awful place, where the energy surrounding her was so thick with hatred she could smell it.

She passed more villas along her route to the front gate. Many of them were built to show outrageous wealth and extravagance. They were empty now except for a few that had been taken over by either a U.S. or some other country's government agency. There were no more than twenty nations from all over the world that were supporting in Iraq since the September 11, 2001 attack on the World Trade Center.

Roxanne stopped to look closely at a bougainvillea bush growing wild along the sidewalk. Bright pink blooms engulfed in glossy green leaves. She stooped and placed her nose close to one to smell its sweetness, but there was none. That's more than I should expect, she thought. A small wind passed momentarily, and she closed her eyes and lifted her sweat-streaked face. Her mind took her home again, to her own backyard in New Orleans where mountains of pink and red bougainvillea bloomed all year long. Her mum loved to garden. And when she was home, Roxanne was drawn to the thick sweetness of the yellow roses that seemed to grow wild there. In the corner of her mother's garden, the bees hummed as they made their way through the honeycomb, small drops of honey dripping silently on the daisies below. Ants moved in a continuous parade between the droppings and their home beneath the old oak tree.

It was lunchtime and she was on her way to Saddam Hussein's presidential palace, a huge marble structure, ten times more outrageous of a show of wealth than any of the villas in "The Canals." The palace was also in the Green Zone. This was where the military who were stationed in Bagdad lived, in trailers on the grounds of Saddam's palace, while the top generals from all involved countries had their separate suite of rooms inside the palace for offices. This was also where all meals were taken—in Saddam's grand ballroom, which was converted to the dining facility. The room had huge soaring marble columns, which held up a domed ceiling of light. Hundreds of hungry military personnel passed through its doors for "three squares a day." This was Roxanne's destination, and the military shuttle bus would take her there.

The small bus stop was about a twenty-minute walk from her villa headquarters outside of the gated community. She approached the gate where the Gurkha soldier stood guard. Gurkhas, she had been told, were fierce warriors from Nepal. They were hired as extra security for their mission. It was common early on in this war to see soldiers from many nations supporting the overall effort. The Gurkhas were responsible for guarding the villa around the clock, and ensuring only approved personnel entered the gated community. This was a community that had only one road in and one out. It was bound on the other side by the Euphrates River.

As she passed the soldier, she noticed his wooden rifle looked ancient, but well cared for. A long-curved bayonet at the end gleamed in the sunlight. The carved wooden stock well-polished. He stood at attention, brown eyes staring straight ahead, holding the weapon directly in front, bayonet pointed up. He had a slim but muscular frame, his uniform clean and neat. He looked young, perhaps in his early twenties. His skin, the color of caramel, glistened in the heat. No stubble appeared on his smooth face. It was not their custom to salute. "Madam," he said as she passed.

"Good afternoon," she said.

"You eat well," he said.

This must be a customary concept that did not translate into English, she thought. She nodded and kept walking.

The bus stop was only a few hundred feet away from the gated entrance, and she surveyed the area as she drew closer. No one was waiting. She patted the side of her cargo pants where she carried her cigarettes, wanting to light up right there. She knew she was breaking a force protection rule by walking alone.

The sun, directly overhead by this time, caused sweat to roll slowly down her back, eventually soaking her underwear that clung to her behind. The thick folds of her battledress uniform scratched at her skin. The bus stop was just a corner on the now deserted street. A small sign read in English "Bus Stop."

Her foot hit the curb as she turned and faced the street to wait for the bus, but suddenly the world became very silent, and she had the sense that she was able to see it with great clarity for the first time. The silence seemed to hang in the air the way the fog would roll in off of Lake Pontchartrain back home in the early morning hours as the sun started to peek its head above the horizon. She remembered what her dad had told her just as she prepared to board the airplane that would take her so far from home. "I'm scared Dad," she had told him, "I don't want to go."

He had looked down at her with such sorrow in his eyes. A Vietnam veteran himself, he had pulled her close holding her against his warm chest. "When it's your time, it's your time," was all he managed to say, and then whispered, "I love you."

Just then, in the distance she heard the start of the noon call for prayer drawing her back. She'd grown used to hearing the adhan sound five times a day, but this time it was as if she were standing just under the tower. She heard the muadhan's voice clearly from the distant mosque's minaret tower ring out, "Allahu Akbar."

She saw the heat rise off the broken asphalt in small flickering waves. *"Ashhadu an la ilaha illa Allah."* Then it happened. She saw a flurry of movement out of the corner of her eye. But she also felt in that moment a sense of déjà vu.

Suddenly, the air turned static, electric. She smelled fear, both hers and theirs, an acrid and metallic smell. It tasted bitter on her tongue. The hair on her neck stood up, and her nostrils flared. Every muscle in her body tightened, as if she were a human spring ready to launch. *"Ashadu anna Muhammadan Rasool Allah."*

The first car sped up on her right side and fully entered her vision. Her head turned to look at them at the same time her hand went to her weapon, a military-issued 9mm Beretta. She pulled it from its holster. She stood on the tips of her toes, her body leaning forward in a running stance. In her ears, the sound of loud pounding. Somewhere she realized in some small part of her brain she was hearing her own heartbeat, but everything else sounded like white noise. A vision of floating above herself occurred at that moment. She heard the noon prayer, a loud melodic voice in the air, *"Hayya 'ala-s-Salah."*

She observed the men in the car screaming at each other, but she was unable to hear their voices. Only the white noise. That pounding was present.

Their bearded faces contorted and convulsed; they argued with each other. There were four of them, two in the front and two in the back. They held some sort of long gun, but all she saw was the rusty nozzle of the weapons sticking up through their clenched hands. Their fingernails are dirty, she thought. The rusted car was riddled with what appeared to be bullet holes all along the side. They turned and stared at her through their dirty windows. The car stopped in front of her, blocking one side of the curb.

"*Hayya 'ala-l-Falah.*"

She had been standing on a triangle of road, two sides facing toward the empty street and behind her a vacant building. There were only two avenues out of there. She calculated the distance to the curb as she sensed another car approaching. Again the feeling of déjà vu welled up inside of her.

The tingling of every muscle pulled at her bones. The sun shining directly overhead. No shadows beneath her feet. The acrid smell. Her mouth like sandpaper. Despite the heat, her sweat ran cold down her back.

Out of the corner of her eye she saw the flicker of the second car entering her space. It was speeding, but for Roxanne, it seemed to be tracking as if it had somehow passed through a time warp causing it to appear in slow motion or as if there was no such thing as time. Every molecule hung in the air like tiny bees in a huge hive—the world was shifting, and she was somehow there inside and outside of it all in that one moment.

As she turned, the second car started to slide into the empty curb in front of her. She looked inside. It was some type of SUV with three more bearded men, two in the front and one in the hatchback. The one in the back held what appeared to be a rusted machine gun. Bullet holes riddled the side of that car too. In the distance, the noonday call to pray continued to ring out. "*Allahu Akbar.*"

On the other side of that triangle, with the arrival of the second car, she saw the opening in front of her closing. At the sight of the first car, her body was already in motion, had been in motion. She was running through that closing gap. She and all her molecules were in motion, sprinting forward. The cars were closing in on that gap, while the gap was closing in on her.

She saw the two cars coming together. The space between them getting smaller and smaller. And in that moment, she was leaping through that gap, or through the cars—she wasn't sure.

Their faces contorted, screaming at her and each other. Their car doors a blur of opening. There was so much movement. The pounding in her ears was unbearable.

Suddenly, without knowing how, she was on the other side of that roadblock of two cars. She could not feel her legs, even to tell if her feet touched the ground. She felt weightless as she ran toward the gate. It felt like flying.

"Gurkha!" She heard a voice scream and realized it was hers. "Gurkha!" There was movement from the Gurkha at the gate. By this time, she had closed the distance, and was nearly halfway there. «Gurkha!... Help. Help," she screamed, lungs on fire.

They are probably following me, she thought, and turned around, arms outstretched, weapon at the ready pointing it downrange toward the men. Without one additional thought or hesitation, she was prepared to pull the trigger and take a life—even her own.

And then time seemed to speed up. She saw them staring at her as she ran backward now, gun in hand aimed at them, ready to shoot.

They looked perplexed, and were still arguing. Perhaps they wondered how one small woman in full battle gear could outrun seven men in 125-degree heat. A mystery—or a miracle? She heard the Gurkha wail something in Nepalese as he stormed from his post, bayoneted rifle at the ready. As the space between her and the Gurkha closed, she turned back around to face him and held out her hands, showing him her weapon, while pointing at the men behind her. His face fell silent as if in meditation—his eyes met hers and locked. A burst shattered the air in front of her. His body hung in the air for an instant before he fell abruptly to the pavement, arms outstretched. A small pool of red spread like roses from his head. A flock of blackbirds sprung from the trees above. She felt a spray of something hitting her back, but for some reason, she could not place the sensation—only that it felt like melting into a sea of asphalt and sand. The last thing she saw was a blur of pinkness as her head hit the ground. Before her eyes closed, she saw one lone pink bougainvillea, its head hanging down mimicking hers.

"La ilaha illa Allah," the final call to prayer reverberated through the noonday sky.

Her body hung in the busy marketplace outside the Green Zone for two days before the special operations soldiers were able to go in by night and retrieve it. Al-Jazeera had captured the video of the marketplace that now played even on the Western media. "A horrible loss," her commander said.

The sun was just starting to streak through the morning sky when the men knocked at her mother's door. Soft colors of pinks and a tinge of orange accompanied them. One wore a cross on his lapel.

Small beads, like tiny diamonds, shuddered on the leaves, while ladybugs swarmed a praying mantis frozen in the grass.

They said, "I'm sorry, ma'am." They wore their dress blues, gloved hands held the envelope of condolence. "I'm sorry," they repeated in unison.

In the backyard the bougainvillea were opening their blooms turning their faces silently to the sun.

Jessica Evans

Half Moon

Winter was the only season we could be together.

I tried to tell you in my care packages sent to that APO address, a weird conglomerate of numbers and letters. Rank, last name, first name. It seemed so impersonal, those address labels of flat rate shipping boxes I filled with things I thought you'd want over there. Boxes of cookies and smashed pretzels; little trinkets from home that reminded me of you. Of us. Over there, that giant sandbox. You called it a theater once, as if you and the rest of those soldiers were just acting, playing Army, rehearsing lines in exclamation over explosions, waiting for the moment when the curtain fell and you could once again be my Suzanne. Your nine-month rotation was a lifetime for me; for you, in makeshift barracks in your forward operating base, the cacophony of mortars sounding off in the distance, it was eternal. I missed you like I never understood could be possible. I imagine it's something like my mother missing her China, but filled with more longing, less truth.

You received your orders at the tail end of summer and we spent those middle months between Labor Day and Groundhog Day nestled between the covers in my basement room, huddled into one another like sybarites, bent on pleasure. Time was our luxury because we knew come 2 March at 0900 hours, you were to report to your duty station, board the bird that would take you across the pond and into the brume of war. Each mark on the calendar left me dying in degrees. Spring Festival, the Chinese New Year, came and went. I couldn't bear to enjoy the celebration. It meant only six more weeks. That was exactly when you started pulling away. Our last goodbye was chaste, less impassioned than I'd spent months imagining. You left for your installation with me full of doubt.

I wanted to tell you I loved you then. But I wanted you to believe it even more. So I kept quiet, silently marking small circles around every day we were able to be together, drawing big lines through the ones we were not. You were my world, but I was just a suggestion in your orbit. Caprice, you moved in and out of my life as if I had no stake in the game. I was indelible. And even still, I sent you something every week, hoping my small offerings might suggest to you that every other woman was and forever will be ersatz.

You couldn't bring yourself to see it. I recognize that truth, now, when it is another summer and you've been back for almost six months but this is only the fourth time we've seen one another. As if I never mattered, as if I were some eschar to will into healing. Algedonic, you. Pleasure and pain wrapped into one gorgeous package that my body can't say no to and my heart wants to love.

Since redeployment, a word that still doesn't make sense to me, no matter how many times you tried to explain it, I'm not sure what your life has become. I do know you refuse to share it with me. I want to go back to winter. Now the world is an open book of possible and there's no telling if we'll ever work; less chance of me knowing what I'm doing with myself. But you're my Suzanne, so I pretend like I don't have these thoughts when you deign to spend moments with me, like now. Finally summer has arrived and with it, you.

We were on the island and I wondered what couldn't this cure. You've got a half-moon of sand stuck to your butt and it looks like a total heartbreak. Lake Sara has never seemed so inviting. We're camping, pretending like we don't have homes. The silence between us all afternoon has been elastic, something that refuses to allow words. They bounce and reverberate against what we know is coming. If I were true, I would know that you are going to leave me. But I pretend like this is just another summer trip, you're just another girl.

We're sitting nearest to the beach, the cabin that I could afford after scrimping and saving and still I know it's not what you want, or what you need. Beyond the slope of the house, near the water, Lake Sara looms. There's a cypress tree in the distance. In fading summer light, it looks majestic. It looks how the color purple feels.

"Start a fire, Chang."

You never call me by my name.

"Now?"

You look at me, wisps of blonde hair falling over your left eye. For once, your hair isn't wrapped in a regulation bun. You've left it loose, and I like it but I won't tell you that. You look young, like an image from a photograph I'd like to stumble on late in life, the kind of silhouette that begs for memory. I'm exhausted by the promises you've never made. "Yeah, now."

"It's July, Suzanne," I all but scream.

All but scream because, like my mother, it seems I can never quite get there. Never can cross the finish line, never achieve the kind of greatness my Bao always tells me I'm destined for. My life has become a series of do-overs and not-quite-all-the-ways. I can't even finish unloading the dishwasher.

"I want to be cozy," you whine, making your voice small and plaintive. It's disgusting and endearing. I could become a deaf man on the silent sounds your voice knows how to make. Your face sours the air between us and grows thick like Cincinnati heat in July.

"Build it your own self." I stomp into darkness.

I stay in semi-darkness for a while, wallowing in the kind of loss that only I can understand. You will never find the words that voice my missing pieces because your family has been here since the Irish started their immigration at the turn of the last century. You're rooted. I know a fire would help us find the kind of closeness we used to share, the closeness from the time before when everything had a meaning and nothing was left to chance. It was a melody I could count on, something I knew the glint of sunlight would create Dopplers against the irises of your eyes. I hear DSO's version of "So Many Roads" in my mind and sway with the melody running through me. The booze helps me move, so I take another long drink of sangria.

The glint of my ring—my grandmother's simple gold band, the one thing that my mother Bao couldn't bear to part with on her voyage from rural China to middle-of-nowhere Illinois, glints against the last of the fading light. Forlorn, lost grandeur against my simple finger. All I've wanted, all I've needed since you came back from the war is that intimacy we once had, the way you could look at me and complete my world.

I never could quite understand exactly what happened in Afghanistan because you would never talk about it. But I know something happened because you left being one way and came back being someone almost the same but not quite. You are a shade of color that's been reduced by the slightest amount. There's some gnawing, gaping hole that you're trying to fill with all of your errant behaviors. Besides, you should be able to tell me everything. Tell me about the long nights, the scorpions, the sandstorms; you should be able to articulate the feeling of your first good meal after getting off that bird for the last time.

Sure, you'd give me anecdotes, the kinds of words you pass at a party, but you still won't tell me why every time you sneeze, your nose drips blood. You refuse to speak about the still-pink scar on your upper left bicep. I've often wondered if your war is the same kind of war that everyone else is fighting—a war on the very sense of self that makes us human.

You haven't been home long enough for me to change my wardrobe for the season and in that time, I've thought we were exclusive but you've found ten other lovers. I've pretended like I didn't care, blamed it on the rotation, the Army, on being a female veteran with little to show for herself. I blamed it on everyone and everything but you. I wanted us to morph into something new, something greater, one solid force that could take on the world and be better for it at the end of the day.

During this entire time, I listened to both your tunes and mine, trying to find a battle rattle that sounded enough like a symphony but left a longing like a good blues song from the 70s. You would periodically leave me and then return, and it was on the return trips that I couldn't help but sing off key. Of course, I haven't said any of this, and you haven't asked. The missing life, those years when I didn't know you existed, months when you stood in a sandbox and I tried to make magic tricks out of life stand like long passages of Hemingway's *The Old Man and the Sea* between us – something to be skimmed, the gist ingested, but the meaning missed.

I breathe in deep and listen for you. At the water's edge, I hear your feet crunching against rocks. What I really want to know is what happened Over There, that giant theater of destruction and regulations, but I can't ever find the right time to ask. And instead of looking for the words, I act out. I need you, but there's no way to interrupt your own grief to talk about us.

I listen hard, wanting to see you rub the lake on your skinny arms, wanting you to want me the way I need to be wanted. At the shore, I swill the rest of my sangria, letting the booze and fruit cloud my head and alter my judgment. Heaving myself from my crouched, typically Asian squat that I never learned but somehow have always known, I wonder if my lips are stained purple. If they're purple, you might notice me, see I'm a person and not another of your platoon, I'm supposed to be your partner, your best friend, I'm supposed to be the someone that you turn to when the sunlight is so bright in your eyes that you need someone to help you see. It's dark, and in the darkness, you cannot see.

Fiction Honorable Mention

Ken McBride

Rose's Blink

Seventeen-year-old Pham Minh Dung lifts the lid to his spider hole and peeks out. It is the fifth straight day of heavy rain and the surrounding area is soaked and gloomy. The Marines have settled in the immediate area and are focused on the enemy across the Song Quan River; they are unaware Minh is on their side of the river and concealed in a tunnel. One Marine, nineteen-year-old Lance Corporal Peter Jasienski, has strayed away from his main group and is only 30 meters away. Minh knows this is an easy shot.

He lifts his AK-47 and points it out toward the Marine, being careful not to reveal his position any more than necessary. Rain splatters in through the opening, misting into Minh's face and running down the exposed part of his rifle. The American, wiping the rain from his face, turns and looks in Minh's direction, his helmet pushed back. The young Viet Cong, being an excellent shot, aims and prepares to squeeze the trigger.

The Lance Corporal's hometown, Cherry Hill, New Jersey, is antipodal to his location in Vietnam and is twelve time zones away. His mother Rose feels the emotional strain of the distance between them. Peter is scheduled to come home within a month and she is jubilant about the thought of his return. She is baking his favorite cookies to be sent to him the next day. The smell of the cookie dough makes her reflect on his early life, remembering when his Little League team won the regional championship, and a few years later when he threw a no-hitter in his senior year at Cherry Hill High School. She remembers how handsome he looked with his date as they left for the prom and his senior trip to New York City. She also remembers when the Camden County Sheriff busted him for drag racing just before leaving for the Marines, and how she made him pay the fine with his own money.

Consumed with these thoughts, she slides the cookie sheet into the oven. As the heat rushes out it causes her to blink.

10:08:22.980 am
Minh pulls the trigger.
The cartridge explodes in the breach and starts the projectile down the 14.5" barrel. It leaves the muzzle at 2,350 ft. per second (715

m/s)—more than double the speed of sound. Moving at supersonic speed the projectile creates a mini sonic boom. Pressure (sound) waves emanate from the breach in concentric patterns.

The force of the explosion transfers kinetic energy to the bullet equal to its mass times its velocity squared, divided by 2, about 2,106 joules. Following Newton's third law of motion, the rifle starts to recoil.

The round is traveling toward the Marine with essentially no loss in speed—it is halfway, approximately 15 meters—and right on track to hit his head above the eyes and slightly left of the bridge of the nose. The temperature of the projectile has risen to approximately the point of boiling water but far below the melting point of lead.

The rain and atmospheric conditions cause the Minute of Angle (MOA) to suffer a slight increase, causing the projectile to strike the supraorbital ridge of the 5mm-thick bone of the Lance Corporal's forehead. Much of its kinetic energy is dispersed instantly and its speed is dampened significantly. Its impact momentum is its mass times its velocity, approximately 5.76 kgm/s, enough to knock him backward but not necessarily off his feet.

The bone starts to crack as the bullet grinds through the forebone in less than .000006195 seconds. Two fissures radiate from the point of entry. The projectile is split into three fragments by the impact. The largest piece stays on the original trajectory, but the two smaller fragments skew downward at 30 and 45 degree angles respectively.

All three fragments enter the frontal lobe and the cortex. Their temperature has dropped to 190 degrees Fahrenheit and the speed has diminished by a factor of three. Much of the kinetic energy was expended breaking through the squama frontalis and the velocity of the fragments is now subsonic. The remaining kinetic energy is apportioned according to their mass.

The largest fragment crashes through the pre-frontal cortex and crosses the central sulcus of Rolando. The Lance Corporal has lost working and episodic memory. He will no longer have memories of high school or boot camp or even yesterday. Planning of complex, coordinated movements is gone. Peter's remarkable ability in throwing a baseball is erased and he will never again tie his shoes.

Continuing into the parietal lobe with some loss of velocity and energy, it starts to yaw and tumble as it moves through the cortical material. The chunk of lead has severely damaged his somatosensory function. The Lance Corporal's muscles, joints, skin, and fascia receptors have fallen silent. He will never again experience the soft skin of his girlfriend or the soothing feel of a warm blanket on a cold night.

The projectile passes ventrally by Brodmann's area 5 and splashes through Brodmann 7, destroying Peter's ability to determine where objects are located in relation to parts of his body. The location of the doorjamb and the bathroom sink are now an unsolvable mystery. It then enters the occipital lobe and eliminates the Marine's ability to process visual stimuli. The light from the outside world is turned off. Peter's ability to experience a vision of his mother's face or his own image in the mirror is gone forever.

The bullet fragment blows a hole in the skull. White and gray matter explode from the opening and splatter in the Marine's helmet. There is no longer enough energy to punch through the steel of the helmet but what energy does remain drives the fragment to traverse orbitally around its interior, dropping out of his helmet above his right eye.

The smaller pieces of lead skew downward from the point of impact. The one traveling on a slightly higher plane plows into the diencephalon destroying the thalamus, hypothalamus, and most of the limbic system. It continues through the midbrain and into Wernicke's area. Peter's capacity for speaking or understanding language is knocked out. He will never utter another word, nor will he again receive or process sounds from the outside world.

The fragment traveling on the lower plane rips through the ventral surface of the frontal lobe, eliminating aspects of the reward system and Peter's ability for flexible and creative thinking. He always loved the smell of his mother's baking cookies but with the destruction of his bulbus olfactorius, this experience for the Marine has happened for the last time. The ventral tegmental area is destroyed, taking with it Peter's ability to experience sexual arousal and pleasure. It slices into the pons where it stops, eliminating communication between upper and lower brain regions, and ends the capacity for slow-wave and REM sleep. The Lance Corporal will never again experience a dream.

The ascending reticular activating system ARAS necessary for a sense of "self" has been destroyed. The Marine will never again have the experience of what we perceive as consciousness. No neural activity in his brain will ever again give rise to the subjective experience of "I" for Peter Jasienski.

The Marine is now beginning to fall. There are no longer projections from the cortex directing motor response or causing the muscles to do anything. At this point the Lance Corporal's body is no longer of any value to him. His helmet is unbuckled and when he hits the ground it comes off in a flood of red, his face pointing upward, and his eyes wide open with a frozen stare.

10:08:23.023 am

Forty-three thousandths of a second has passed and Rose's blink has yet to finish. As her eye opens her mind floods with thoughts of Peter's return as she envisions the future: TWA flight 597 lands on time at Philadelphia's International Airport. As it taxis to the gate, Rose feels a surge of excitement. The mechanized stairs roll up to the plane and the cabin crew opens the door. She watches as the passengers disembark and then sees Peter. He is waving and wearing his Marine uniform. She thinks he looks thin but otherwise okay. Soon they embrace and hug each other. Rose has been longing for this moment ever since he left. She cannot wait to get home and cook him a big meal along with his favorite cookies. Once home Peter tells her about Vietnam and the things he encountered. He asks about his friends and his girlfriend. She tells him the family all want to see him and welcome him home, so they will be coming over to share a big meal. She is excited to show him a letter that arrived in the last week from Lehigh University saying that his application has been accepted and that he can enroll in the upcoming fall semester after his discharge from the service. Rose has been looking forward to this for the entire time Peter has been gone.

Three days later, a knock at the door. She answers to see two Marines in dress blue uniforms.

Breanne M. Pye

Stitch

Tuesday is no different from the six other miserable days of our deployment routine. We wake up drowsy after a night of frequently-interrupted sleep, and stumble out of the dark of our tents to be assaulted by all of our least favorite things. The sun is the earliest riser of all in Afghanistan. It blisters our skin from the moment we wake until we slink back to our tents to put the terrors of the day behind us. So too does the earth assault. The first steps off our tent platforms onto wooden boardwalks result in a cloud of dust that creeps through even the tightest-laced combat boots and creates a second skin of sweat and dirt that we will be forced to wear for the remainder of our time in this unforgiving landscape. But we are soldiers, so we wake, scrub our second skins the best we are able, and attempt to look the day in the face, resolving to carry on with the business of war. I say "we" like I actually belong in this collection of indifferent machines disguised as human—but I'm no warrior; I'm just a uniformed witness to a dark chapter of history, one which both began and ended on a Tuesday.

Walking to the chow hall requires excessive concentration. The road is a thick bog of unsettled gravel and I have learned the hard way that ankles are fragile things. I've also learned never to look too far ahead of myself while walking. The heat collected by the gravel creates a wavering veil that distorts every detail of the landscape and instantly turns my stomach sour. I avert my eyes as I wade through, carefully navigating an army of stray cats which are here because they are attracted to the vast assembly of waste that only the privileged can accumulate. As I approach the chow hall, I hear a commotion coming from our Role 3 medical clinic so I take the dirt road to the left instead of the gravel road to the right to go see what all the fuss is about. I welcome any distraction from this day, which has already lasted thirteen months. Turning the corner and looking ahead, I'm greeted by the heat-distorted image of two young men in digital uniforms carrying a heavy plastic-wrapped package toward a waiting truck. A few steps closer and I hear laughter as the men dump their package into the truck bed.

I wait until the soldiers leave before heading in for a closer look. As they walk away, their boots kick up dense clouds and I can hear them laughing as they chant "and another one bites the dust." A chuckle catches in my throat as the cloud dissipates and I get a clearer view of the package they've just deposited into the truck bed. A steady stream of red leaks out of the torn corner of a black trash bag, carving a path through the dirt of the lowered tailgate before crawling over the edge and creating a new cloud as it hollows out a bowl in the dust below the truck. I hesitate before lifting the edge of the bag up for a peak. The chuckle suddenly feels like a hand tightening around my throat as the source of the bloody stream comes into focus.

Coarse black hair coats the brown flesh of what I now recognize to be a human arm. My hands shake and my knees begin to lose their strength. My eyes follow the blood trail in slow motion up from the crater below the truck, back over the tailgate, and through the black hair until it disappears beneath the edges of the trash bag. It's not the blood that finally takes my legs out from under me—it's the arm with no hand attached to it, just a rounded bit of scarred flesh where the hand used to be. My choke turns into a cry and a tear carves its way down my cheek before mingling with the red bowl beneath the truck.

I know this arm.

*

A Lockheed C-130 Hercules aircraft looks like a fortress when you're standing in line to board it. But after countless hours cramped together with dozens of soldiers within its belly, the plane feels like a dungeon. I landed at Kandahar Airfield on the first Tuesday in July— the start of my deployment. The rear ramp lowered with a slowness to which only machines are indifferent. The sun's first attack was met by a mass of disoriented soldiers as the ramp thudded against the tarmac, exposing me to heat which threatened to scorch me from the inside out as I inhaled it for the first time. The impact immediately caused anyone with hesitant knees to crumble into a pile of bones over which I stepped in an attempt to escape.

Finally able to disembark, our training took over and we sorted ourselves into ranks of fifteen while medics hastened up the ramp behind us to attend to those who had collapsed. A line of short, brown-skinned men clad in off-white *khet partug* tunics walked through the heat-veil and greeted our commander, who growled at us to make our-selves presentable. As we waited for an introduction to the strangers, I was reminded of pictures of the Israelites in the Bible stories my father used to read to me every Sunday night of my childhood. These men could have walked straight out of those pages, from their blue turbans to their *perahan tunban* trousers and hemp-rope sandals.

To my left, our youngest private elbowed me in the ribs, or shoulder rather, as I was a good foot shorter than anyone else in my squad. "It's rude to stare, Sar'nt" he hissed down at me.

"You're the last person I'd expect to be schooling me on manners, Murphy," I whispered back as our commander began assigning the Bible men to each squad.

170

"Listen up, boys!" he barked. Yet another reminder that I was sorely out of place in this all-male infantry platoon. "These men will serve as your interpreters this deployment. They tell you to do somthin', you do it. No lip!"

One of the Bible men stepped up to our squad leader to introduce himself. I glanced at him out of the corner of my eye and noticed he was wearing a cream-colored *pakol* cap instead of the traditional turban donned by the rest of his fellows. The cap rested low on his brow and framed deep lines that crinkled around his eyes when he smiled.

"Welcome to Kandahar," the man said, his eyes squinting closed as he chuckled. "I'm sure you'll love it just as much as I do by the time you leave." There was humor in the man's voice, but none of the boys seemed to catch it as he continued. "I am Azad. Roughly, this means 'free' in Dari, yet here I am. I think you call this 'irony' in America."

I laughed out loud at that, drawing a glare from my squad leader and a curious glance from Azad. "A woman?" He asked, walking over to stand in front of me. "This is very unusual." With that, he offered me his left hand to shake, setting off alarm bells in my head, as I had been trained that it was offensive for an Afghan to touch another person with any part of the left hand. I wondered if he was insulting me because I was a woman, but the thought drifted away when Azad lifted his right arm to his chest and I noticed that it ended just above the wrist in a lumpy bit of scarred flesh which was much lighter than the rest of his skin. I shifted my M4 rifle strap from my left shoulder to my right so I could offer him my left hand without pointing my rifle barrel in his direction.

"She's our photographer," my squad leader explained, "here to take photos for historical documentation and whatever the fuck else they do." Next to me, Murphy snickered.

"Combat camerawoman, actually," I corrected my sergeant, offended by his apologetic tone and obvious embarrassment. "Pleasure to meet you, Azad."

Azad squeezed my hand, meeting my eye. "I've never seen eyes as blue as yours," he almost whispered. "It will be a delight to see such a bright spot of color in this brown sea for a while."

Friendly Azad and his missing hand. Right away the boys refused to trust him, refused to include him, refused even to acknowledge him. After our exchange, Azad moved down the line of soldiers, continuing with his introductions. As he offered his left hand to Murphy, the private jerked his hand back at the last second, spitting on the ground at Azad's feet. The spit sizzled and evaporated as it hit the tarmac, evaporating in a matter of seconds. The boys hardly contained their snickers as our squad leader cuffed Murphy on the head. "Congratulations,

Murphy! You've won yourself a personal assistant position for the rest of this deployment. From now on, Azad had better not so much as take a shit without you offering to wipe his ass!" A deep blush crept up Murphy's neck and settled into his cheeks as his eyes narrowed.

"Yes, Sar'nt," he mumbled under his breath.

"What's that, Murphy?" our squad leader barked, stepping so close to Murphy's face that the brim of his hat bumped against the private's forehead.

"YES, SAR'NT!" Saliva sprayed out of Murphy's mouth as he screamed the response, which echoed down the tarmac and bounced off the concrete wall of the terminal before creeping back to us.

We all stood with our eyes glued dead ahead, trying our best to ignore the tension of the exchange. Azad barely missed a step, he just nodded at Murphy, then at our squad leader, and then he offered his left hand to the next soldier in line. The handshakes were noticeably stiff and unfriendly after that.

I understood the boys' nervousness. It was born from the depths of a persistent dread that all soldiers carry from the moment we receive orders to deploy. From the second we kiss our loved-ones goodbye to the day we make it back home safe, *fear* is the only constant. Long before we ever get on a plane to begin the long journey to war, we are taught to fear the Taliban. We are inundated with training videos and briefings of their terrifying acts and taught that we are the Afghan people's only hope of liberation. And though we know in our hearts that most Afghans are not members of the Taliban, we can't seem to separate the monsters in those videos and briefings from the men, women, and children we encounter when our boots hit the ground in Afghanistan. The sight of a turban or a *burqa* or even a *pakol* instantly brings those videos to our minds and triggers the kind of fear we've been trained to trust.

We finally made it to our combat outpost around supper time. Azad showed us around tent city, where we would be sleeping for the next year and a half, then led us toward the chow hall, giving us tips along the way about how to navigate the loose gravel roads without breaking an ankle. The boys were quiet, mostly ignoring Azad. As we made our way through clouds of dust that lingered above the gravel, one of them coughed and cursed whatever gods had landed us in this hell.

"Yes, yes, lots of dirt here," Azad offered. "But no bugs! This is good, no?"

"Fucking towelhead," Murphy muttered behind me.

I turned around to look at him. "You know towelheads refer to folks who wear turbans, right?" I jerked my head toward Azad. "He's obviously wearing a *pakol*."

The soldiers to either side of us chuckled as Murphy pressed his lips together and glared at me. We kept walking.

Once we arrived, the boys immediately grabbed their trays of food and crowded around one table, leaving Azad and I standing together in involuntary solidarity. Once the two of us had filled our trays, we turned to join our squad only to discover that they had spread their gear out to ensure we'd get the hint: there was no room at that table for a crippled Afghan and a female soldier.

I was used to this kind of treatment by now and Azad didn't even seem to notice it. He waited for me to get a tray of food, then nodded his head toward an empty table in the back of the chow hall. As we made our way there, our squad leader picked Murphy's tray up off the table the boys shared and stomped after us. He slammed it down on our table without a word and pointed at the bench for Murphy to sit. Murphy gathered up his gear and stormed out of the building.

"Eventually, he'll get hungry enough to eat with you," the sergeant said before turning to rejoin the boys at their table.

Azad waited until the sergeant was out of earshot before retorting, "I cannot wait." He winked at me before picking up his fork to eat.

"How can you stay so *calm* when we treat you like this?" I asked.

"*You* don't treat me like this," he answered. "It is enough. Those boys? They haven't seen war yet. They will come around. You will see."

Eventually, Murphy did get hungry enough to join us in the back of the chow hall, but he sat on the opposite end of the table from us, regularly pulling out his sketch pad and working on charcoal caricatures as he ate, doing his best to pretend Azad and I didn't exist. Azad made every effort to include the private in our discussions, frequently bringing him new charcoal pencils from the *haji* shops in Kandahar. They must have cost him a small fortune, but Murphy just shoved them into his pockets with little ceremony, often making comments like "I suppose that's the best I can hope for in this shithole."

Over the course of the next several months of our never-ending deployment, the boys continued to ignore not just Azad and I, but Murphy as well. Azad served as both escort and translator when I joined our command staff in village *shuras*, where it was my job to document all the "progress" we were making with the elders of Kandahar Province. During these meetings, Murphy was forced to act as a runner between Azad and the elders, while the boys maintained a security perimeter outside the *shura* location. Their resentment for both Murphy and I grew as we were always invited to partake in lavish Afghan feasts with our command staff and the Afghan elders while they baked in the sun outside and were forced to eat MREs. Though Murphy complained as often as any of the boys would listen, they took to calling him "The Shadow" and goaded him for being "Azad's bitch."

It didn't help matters at all that the boys were limited to building security perimeters for peace-keeping missions that were held on secured Forward Operating Bases or heavily fortified palaces within the city. They grumbled about having to serve as bodyguards while the rest of the men in our unit were out getting "action." Too frequently, they saw Azad, Murphy and I return from the gruesome scene of Taliban attacks where Azad served as a translator between myself and the Afghan National Police while I documented the scenes and Murphy served as a runner. The boys all saw the blood on our boots and the haunted look in our eyes when we got back to our combat outpost—but instead of offering us their comradery, they continued to isolate us and take their resentment for the "action" we were seeing out on Private Murphy, who became more and more reserved as the deployment progressed.

As days threaded into weeks, then months, I found myself increasingly thankful for Azad's steadfast presence and especially his empathy and kindness in helping me process the grittier aspects of war. Though I knew he was no stranger to the scenes we'd been documenting for nearly a year, I often wondered how he handled it all with such grace and fortitude.

One day we were sitting at our table in the chow hall after returning to base from a particularly bloody mission. Azad sat across from me, watching me pick at my food without eating anything, trying his best to get me to meet his gaze.

"You must eat," he admonished me. "You cannot keep taking pictures for hours in this sun without eating."

"How can you eat after what we just saw, Azad?"

"I eat because I need nourishment," he said, his voice suddenly soft. "And because I can't expect a next meal and a next meal and a next meal. This meal is right here. It is enough."

"But how do you swallow the food?" I asked, tears stinging. "How do you breathe?"

Azad closed his eyes and took a deep breath before opening them again and moving our trays to the side. Then he reached across the table with his left hand, palm facing up, inviting me to take his hand in mine, which I did. He squeezed it and met my eyes, a familiar gesture by now, holding his scarred right arm up for me to see.

"Today, I will tell you how I lost this hand. Then, we will breathe."

"I was born in a *Kochyan* village," he began. "These villages are everywhere in the north, where our nomadic tribes spend their

174

summers. You maybe see them when we go out in the country for *shuras*. I think you call them 'Kuchi' camps. And so, in the *Kochyan* villages, we don't name our children until their fifth birthday. The little ones are vulnerable. They get sick easy and can't handle life on the trails so well. Many die. Many are taken by the Taliban to be raised as warriors *insha'Allah*—if God wills it. My brothers, they were both taken before I was born. I was not taken, so they named me *Azad*—'the free one.'"

With this, Azad paused, letting go of my hand and rubbing his right arm as he took another deep breath. He closed his eyes, placing both of his arms in his lap, as if he were trying to gather the warmth before continuing with his story.

"And so, I help my parents and my uncles with the sheep and goats. When the sun goes far away from the earth and our tents can't keep out the cold, we take the family and the animals over the mountain passes. Here, into the southern provinces. Because my brothers fight for the Taliban, we don't pay the tax to cross the mountains. We will be cared for and my sisters will be spared becoming brides of *jihad*, *insha'Allah*. This is a great joke. The Taliban always needs brides and *Allah* always wills it to be so.

"And so, my fourteenth summer in Kandahar, we are setting up our village in the Arghandab River Valley where there is plenty of water for drinking and washing. Every month, the Taliban comes to collect the rent for using the land: one of every five chickens; one of every ten goats; one of every twenty sheep, and so on. It is very expensive, this rent. And so, three days after they come to collect the rent, I am standing in the yard, sweeping the dirt, shooing the chickens away from our tent. I hear a sound like thunder and I look up, and there are two men dressed all in white on the motorbikes of the Taliban. They crash right through the gate and skid their tires so that dirt flies all over our tent. My *Baba*—my father—he pushes aside the curtain to our hut and comes outside to stand before the men. They tell him to go back inside and get my sisters and to line them up outside for examination. He does this thing. He does not say a word, but he looks at me in a way that I know means to stand very still and try to be invisible.

"When *Baba* gets my three sisters outside, one of the men pushes him aside and begins pinching and squeezing them with his fat fingers while the other man goes inside to fetch my mother. My sisters stand still, looking at the ground, trying to be brave while the man leaves smudges on their *burqas* with his grease-smeared fingers. Inside, I can hear my mother screaming."

I reach my hands across the table to offer Azad comfort, but instead, his eyes crinkle as he opens them and smiles. He takes a deep breath and reaches out with his left hand to pat my outstretched hands, comforting *me* instead. Out of the corner of my eye, I see Murphy turn away from us, his shoulders heaving as he exhales heavily.

"And so . . . " Azad continues, his voice so soft now that I have to lean forward to hear him. "The man outside, he grabs my oldest sister by the back of her *hijab* and he drags her. Her feet kick up clouds of dust as he drags her to his bike, calling for his fellow to come back outside so they can leave. I can not hear my mother screaming now. The man holding my sister, he doesn't even see me until I step between him and his motorbike. I was very good at being invisible, but now I don't know what else to do. I just know he can't take my sister. The man, he lets my sister go and grabs his rifle off the back of his motorbike. He slams it into my cheek. Everything is black and I fall down. When I can see again, the man has me chained to a tractor tire next to our hut. The chain is around my right wrist. The man pulls a machete from a loop at his waist and grabs my chain with one hand, straightening my arm out.

"'We come to bring honor to this family by choosing a bride of *Allah* for one of our warriors,' the man says. 'Instead, this *boy* has interfered in the work of *Allah*.' The way he says 'boy' it sounds like a bad word. He looks at me, then at my family standing in the yard with their heads bowed. 'Those who inhibit the work of *Allah* will suffer his wrath, *insha'Allah*,' he says. Then he swings the machete down. I feel my arm jerk free of the chain. First, I think he missed. Then I see my hand hanging from the tire. The man, he leaves me there on the ground and gets on his bike with my sister. They ride away. Just like that. I look down at my arm. I remember thinking how pretty the blood is as it pours out of my arm and carves its way through the dirt. It would make such a pretty color for our hut. The blackness, it is coming again. I look over at my *Baba*, who has turned my sisters away from me. He looks at me. He is crying, but he does not move to help. I think he is afraid the men will come back. He closes his eyes and turns his head away. I take a breath and pull my arm free of the chain. I stick it in the dirt to stop the bleeding. I take another breath."

As he finished his story, Azad looked down at his missing right hand. "I knew then I could never be a peaceful *Kochyan*, walking the roads of the seasons and worrying about only my chicken and sheep. Nothing will change if we keep blaming all this violence on the will of *Allah*." He looked over at Murphy, whose back was still turned to us, before speaking again. "I have no family left for the Taliban to take. I have no hand for swinging swords or pulling triggers. I have only these two languages to fight with. That is how I breathe."

176

He smiled as he grabbed my tray of food and pushed it back across the table at me.

*

The stream of red gets thicker as the truck carrying my friend's body begins to inch forward. It leaves a trail of dust which follows the truck like a funeral procession. The dust clings to my cheeks and works its way under my uniform to thicken my second skin as I find my feet and step after the truck, which is slowly making its way toward the massive burn pits at the west side of our outpost. I can do nothing but follow behind, suffer the sun's continued assault, and try to breathe.

Two days from now, our commander will gather our squad together in the deceptive shade of a concrete bunker and tell us that Private Murphy has been taken into custody. He will tell us that Murphy will be transported under armed escort to the United States, where he will stand trial for murdering Azad. He will relate these details in a factual, detached manner. He will tell us that Murphy was on guard duty in one of the outpost's security towers when Azad came to offer him a plate of jasmine rice. He will tell us that, somehow, the rice was spilled on a charcoal-sketch Murphy was working on. He will tell us that our youngest private calmly pulled his 9mm pistol out of its holster and shot Azad in the head, with no warning. He will tell us the local national population was afraid of retaliation if they allowed Azad to be interred in their traditional burial plots. He will tell us that our resources are too limited to be able to afford a convoy to bury Azad somewhere out of the Taliban's reach. The boys will chug Gatorade and poke at the dirt with their rifles as they receive the news. They will remain silent but the discontent on their faces will tell me they are secretly cursing Azad a final time for causing them to roast in the concrete bunker. I'll feel my own anger begin to simmer as I recall the last image I have of my friend.

The truck I have been following finally rolls to a stop before the burn pits and I observe that the river of blood has dwindled to a slow drip. No one questions my presence or the presence of my squad, who frame the edges of the pit, their bodies casting long shadows across my face as I walk up behind them, the stench of burning trash in my nostrils. I watch two soldiers pull Azad's body out of the back of the truck. They remove him from the trash bag. Only then, I notice the entire right side of my friend's face is missing. I fall to my knees, retching, the liquid kicking up a cloud that works its way under the collar of my uniform and adds another layer to my second skin. I feel the heat of the blaze rage against my face as the soldiers toss what is left of Azad into the pit. As the flames catch the edge of his blood-stained tunic, the boys disperse.

Cindy McDermott

I Wonder How She's Doing

We sit in her small kitchen in a small farmhouse in small-town U.S.A. Her two daughters play ring-around-the-rosy in the adjoining living room, too young to understand our serious talk.

Her table is covered in a homespun yellow-and-white-checked oil-cloth. The touch is sticky yet slick in the worn spots.

I sit next to her. I hadn't planned it, but women tend to congregate in a protective shell, offering comfort, all the while thankful we aren't receiving the news.

I place my cover on the table and smooth strands of wayward hair that had escaped from underneath it. The gold band at its base and braiding on the brim glimmer. It's the perfect topping for our white uniforms. While civilians love them, sailors hate them. Dirt magnets. Ice cream man suits. More akin to handing out drumsticks and snow cones, greeted by excited kids and anxious parents, digging for change in their pockets.

But today, no one is happy to see us.

*

The night before, I ready my uniform, pressing knife-sharp creases, pinning on my ribbons and adjusting other knickknacks that always catch on my seat belt. My uniform denotes an experienced officer, higher on the command chain. But my queasy stomach signals how unprepared I feel for this mission. Yet I will look my best while delivering the worst.

*

As we sit around her table, CACO, *Casualty Assistance Calls Officer*, the official representative of the Secretary of the Navy speaks to my widow. He reviews funeral plans and details on how her husband's body will arrive. She receives the specifics. So courteous, so polite, so numb. The phone rings. Probably another neighbor wanting details. A family friend, a Vietnam Vet, leaves the table to answer.

News travels fast in a small town. Friends and family would soon arrive, bringing tuna noodle casseroles and homemade pies, giving a different meaning to the words "comfort food." Nosy pokes phone the media, alerting reporters, who call me for verification.

"I can't confirm anything. DOD guidelines are 48 hours to notify primary and secondary next of kin. Watch the website." There's nothing to hide. Just nothing I can say. Everybody knows he's dead, but the computer hasn't made it official.

"Ma'am, would you have a close family friend, pastor or an attorney who could speak for you?" I ask. "They can act as your spokesperson for the media."

My widow looks at me with swollen eyes. "Yes, our pastor could do that."

"You might want to turn off the TV. The news will be on this evening, and they'll probably have announcements throughout the day about his—" I feel my throat catch. I draw in a breath to stay in control. My mind recalls hallway hellos with my fellow sailor during drill weekends. I didn't know him well, but enough that I grieve for him too. My widow pats my hand. She comforts the one sent to comfort.

CACO carries on with another detail. "You and the children will be entitled to survivor benefits."

She sniffs. "They deliberately gave him orders for one less day so we wouldn't get benefits. He said his country was screwing him and a lot of other shipmates."

I glance across the table to CACO. Pained expressions cross our faces. Only orders with 180 days receive full benefits. We're embarrassed by our country's most dangerous weapon: sharpened pencils, slicing budgets, intentionally denying those who serve.

The spark's been lit. Anger in her voice grows. "Now, I have no husband and my little girls have no father, and you talk about the benefits we'll get. Where were you a month ago when he left us?"

My widow begins to sob. I have no defense to offer because there is none. Our leaders have sacrificed morals for money. Besides, we've been trained to never argue with a widow. You take the anger, the hatred, the cries and the agony. It's our mission of the day.

After an hour of discussions, my widow's clipped answers tell me she's overloaded with details and grief. CACO asks if there are any questions. Queries about a future she can't begin to comprehend. He promises to be back the next day to help her through the process. We depart, followed by the Vietnam Vet.

"How was he killed?" he asks.

CACO exhales. "His Humvee rolled over an IED."

"At least he didn't burn to—" His mouth begins to tremble. He averts his gaze to the cornfield across the road.

CACO fills the silence. "I want you to know he volunteered for the convoy."

The vet scoffs and lasers a stare at CACO. This battle-hardened Marine knows the truth. CACO and I pass guilty looks between us. We've just fed him the go-to line, designed to help families deal with their loss. A standard message the military pulls out when no other words bring comfort.

But what could CACO say?

He knew what he was doing?

In defense of his country?

He died a hero?

What military member goes into war, longing for death to become a hero? That only happens in movies, glamorizing war and boasting of patriotism. Packed with sugary words glorifying love of country, flags and apple pie in an attempt to cover up war's horrific bitterness.

The ideal is to come back. Not in a box. Not with pieces missing. Not with nightmares that blast you awake with your own screams. Not with guilt so heavy relief comes by eating your gun.

The truth is, you never come back the same.

*

I shake my head and exhale. I'm in the present, in front of my bathroom mirror, brushing my teeth. Yet another time I've lost myself. When memories of that day more than 15 years ago slip into my simple tasks. No trigger. No control.

I've lost touch with CACO and my widow. I wonder how she's doing. Then again, the outcomes might be too much for both of us. Wounds that shouldn't be opened. Mine so insignificant when compared to hers.

The faucet water runs over the bristles, rinsing out the leftover toothpaste. I tap the brush on the sink to remove the excess. The drops meld together, forming an escape route to the drain. If only they would carry my haunting memories with them.

I'm Alright

"It's alright, it's alright,"
"It's alright, it's OK."

This cadence echoed through my mind as I lay with the right side of my face flattened to the hot sand. My breath coming quick, I almost choked on the dust swirling around me. I couldn't move. I was temporarily frozen on that spot in the desert.

"It's alright, it's alright,"
"It's alright, it's OK."

This was my drill sergeant's favorite cadence and with his booming yet melodic voice, the chorus stuck and played over and over in my head until it became a sort of mantra. Now, so many years after those zombie days of boot camp, I still hold onto that mantra as a lifeline.

"I'm alright," I thought, "I'm OK."

I blew out a breath and watched the dirt rush away like water down a stream, forming a small divot in front of my mouth. I looked at the dark, pebble-ridden earth around me and realized I wasn't in that spot in the desert after all. No. There were different sounds as well. That day on the sand there was a loud blast from the rocket-propelled grenade that had exploded near me—numb silence, then ringing. Today I heard the sounds of teenage children laughing and yelling. I suddenly remembered where I was. I was at my nephew's birthday party and we were all suited up playing paintball. An air canister in the supply shed had fallen and made that loud, explosive noise earlier. I had hit the ground, knowing another RPG had exploded. It hadn't. It had just been a canister and yet somehow that small object had sent me, like Marty Mcfly, back in time.

I pushed my body up and looked around. I was behind a makeshift bunker, very different from the concrete ones found in Afghanistan. This bunker would ward off only paint, not bullets or shrapnel. The splintered walls looked as if an MRE bomb laced with Skittles had been set off right in front of it. The colors were a bit comforting; dizzying and yet grounding at the same time.

I first heard the heavy pounding of the dirt and grass before I saw my nephew sprinting across the field, paintballs flying as he bolted through the crossfire. A few of his friends ran behind him, trying to shoot the assassins hiding behind the walls. My nephew grasped the blue flag and, just as nimbly, darted back through the center of the war zone. This technique usually worked only in movies but apparently in a paintball war with inexperienced players, rental guns, and suitable distraction of the enemy, it worked.

"Ahh!" I heard one of his friends yell, "I've been hit!"

The bright yellow smear of the busted paintball was evident across his clear face mask.

"Ugh, me too," called the second friend who raised both hands up and began walking off the field with his marked brethren.

This left my nephew, still flying like a bat outta hell, with all guns aimed at him.

"It's alright, it's alright,"

"It's alright, it's OK."

Sometimes, after you come back, you have to remember to have fun. Others can't always see the turmoil going on within you as you adjust to the difference of life back home. Most times it's on you to recognize.

"It's alright, it's alright,"

"It's alright, it's OK

This was just a game. Nobody had noticed me lying prone on the dirt for those few seconds or minutes. They had been busy trying to capture the flag.

"It's alright, it's alright,"

"It's alright, it's OK."

I couldn't leave my nephew out there alone. I jumped up and ran to the center, a move I knew would get me littered with paintballs but I didn't care. I covered him from the rear, providing suppressive fire. Somehow we both made it through the gauntlet unscathed.

When he reached our team's side of the fence, all his friends cheered as he threw the flag in the air and proceeded to do his own version of a touchdown dance.

I laughed and watched as the team hollered and talked smack. Just a bunch of kids out here having fun.

"It's alright, it's alright,"

"It's alright, it's OK."

My nephew's broad smile was turned on me as he gave me a bro hug and I realized for the second time that day, and for many days to follow:

I'm alright. I'm OK.

No Room for Soft

Protocols

Pearson looks over the zig-zag entry into FOB Salerno, his M-60 perched atop sandbags like the bunched body of a black panther ready to pounce. Same posture he should be in, he knows, but he's zoning, his mind ten thousand klicks from desert and guard duty. Mail call this morning brought a care package from Mama—corn fritters, sweet potato pie, tubs of dirty rice sprinkled with corn and pecans and carrots—and all he can think of is chowing down at shift change, his mouth watering at the prospect.

Below the aim of his barrel, Jackson stands at the boom barrier in a Kevlar-plated vest, his torso as blocky as a Lego figurine. He's supposed to ID and search any Afghans while Pearson provides overwatch. But it's Thanksgiving, foot traffic's light, and the unit is rotating home at the end of the week. The rest of Task Force Devil is squirreled away within the HESCO barriers and tank ditches surrounding the compound, eating holiday dinner and, just like Pearson, thinking of home.

Pearson shakes his helmeted head to focus on his task. Seeing this, Jackson laughs then touches his fingers to his lips and blows him an air-kiss. Jackson is Pearson's best bud. All those patrols. All those raids with kicked-down doors and concussive waves of flash-bangs. All those inside jokes signaled by no more than a glance. This morning, over powdered eggs at the DFAC, he'd asked Jackson what he'll do with his combat pay once they get home. *Cruise the strip clubs on Yadkin*, he'd said. *Make it rain on the honeys.* Pearson had fist-bumped and laughed, elaborating on what he'd also like to do with the strippers. But really, all he wants is to sit down to a home-cooked dinner and fall asleep in his old bed, unclenched, far from mortar rounds and rocket attacks and callouts to scenes where someone's gotten their shit blown off.

Blinking back to the present, Pearson sees a woman limping toward the gate. In years to come, reflecting on this moment, he'll wonder if something in the woman's gait reminded Jackson of his crippled mother, the way she caned her way into his bedroom at night to kiss his forehead. Or if, framed within the woman's purple hijab, her expression, angelic as the Virgin Mary, prompted memories of Baptist sermons, echoes of *Thou shalt not* staying his trigger finger. Protocols existed for situations like this, Rules of Engagement outlining the rapid progression of escalating force to use against anyone approaching the gate. *Shout. Show your weapon. Shoot to warn. Shoot to kill.*

But he'll never know for sure why Jackson stands motionless and lets the woman come unchallenged, shuffling toward the boom barrier, hugging a bundle the size of a swaddled baby.

When Pearson's fog finally lifts, he yells *Stop* at the limping woman. Hair bristles at the back of his neck as he notices a mottled bruise swallowing one side of the woman's swollen, fearful face. He fires a warning shot over her head and lowers the aim of his barrel. But before he can fire again, a fist of sand knocks the air from his lungs and the world goes brown.

Visualization

In Pearson's dream, Jackson is a statue looming over his own grave, his cement skin fissured with green mold. Out of a swirling fog, the limping woman emerges, a bloody fetus cradled in her arms. His mind's eye zooms in on her bruised face. *Your mama wanted you to have this*, she says, holding out the stillborn child with wires protruding from its stomach.

Pearson wakes in a sweat, his face and neck throbbing. He touches his cheek, feels gauze, and comes back to himself, looking about. He's in a containerized unit, a rack-slung IV bag at his bedside drip-drip-dripping its contents into his veins. Bandaged and moaning bodies fill the five other beds. The reinforced walls are strung with Christmas lights and painted with a symbol of a sashed cross atop a rose. Scrolled text above the symbol reads "399th Combat Support Hospital." On a radio somewhere, Mariah Carey croons "All I Want for Christmas is You."

Seeing Pearson awake, a male nurse steps over and checks his chart. "Good to see you awake," he says. "Your next dose isn't for another half hour, but I can give it early if your meds are wearing off."

Pearson tries to respond but his tongue feels thick. "I'm all right," he croaks. "How long?" He means to say more but that's all he can manage.

"Since yesterday. Save your voice." The nurse grabs a capped cup with a straw poking out the top and holds it up to his mouth.

Pearson takes a greedy sip. Just water, he knows, but it tastes sour as vinegar right now. The nurse sets the cup on a nearby tray table then moves on to the next bed.

Long ago, when Pearson had been suspended from middle school for fighting, his mama taught him a calming visualization technique, focusing his anger onto a fat Buddha whose serene smile grew larger the harder he squeezed his belly. Whenever he felt like hitting someone, he was supposed to imagine this symbolic figure unfazed by trouble.

Pearson employs this technique now, imagining his hands vising the Buddha. But the expanding, envisioned mouth seems more a sneer than a smile, and the tense muscles in his body tighten further instead of easing up, a red tunnel engulfing his mind, constricting like a python.

A voice in his head says, *You ain't dying here.* But it's not his mother. This is pure Jackson, his honeyed bass rich and deep as a radio deejay's. *Uh uh*, he says again. *Ain't dying here. Gonna get back in the shit. Gonna get your payback.*

Pearson closes his eyes and pictures the smiling Buddha coming to life, six feet tall, kicking in doors. In one dark room after another, robed men snarl like caged animals, lit by the strobing flash of his muzzle. A cold resolve settles in his bones, the closest he's felt to calm in months.

Light Duty

First day back from the CSH, Pearson tells everyone he's fine. "A few scrapes, some stitches, that's all," he says. He lobbies Sergeant Payne to put him back on the line, back to tight alleys with sandstone walls, back to blasting locks and kicking doors. But the Army has protocols for reintegrating wounded soldiers. Assessments, checklists, rotations of light duty in the rear.

So he sits at a desk in a cold, concrete building surrounded by keyboard-tapping POGs who've never stepped foot out of the wire. His temporary job is "Runner" for brigade staff. The tent his four-man fire team shares is the size of a one-car garage, cots pushed against one another, sweaty clothes drying on 550-cord strung between posts. But this office is cavernous and air-conditioned, no sand in sight. When the phone rings, his job is to scrawl a message and run the pink scrap of paper across gleaming floors to the appropriate officer's aide. His phone seldom rings. And when it does Pearson can tell from the aides' curdled expressions that anything coming through CQ is not important enough to pass along to their bosses. If it were, the call would have come in on their lines.

Mindlessly rubbing the scabbed-over pits and scrapes on the left side of his face, he stares into space, cabinets and inboxes and coffee machine blurring into a haze. And from this fuzzy cloud, a familiar face materializes.

Chicks dig scars, Jackson says casual as ever, shuck in his step. He sits down and props his feet up on the desk, bloused bottoms of his BDUs tucked into spit-shined Corcorans.

Just a daydream, Pearson knows, but man, it seems so real. The flattened nose, the pencil moustache, the high cheekbones, the half-lidded look he always gave when talking about getting stoned or getting laid. *What you smiling about?* Pearson says in his head. *Don't you know you're dead?*

Shit, Jackson says, waggling his head to mimic their asshole of a platoon sergeant, Stackhouse. *Don't go soft on me. Man up, Bitch.*

Pearson feels tension seep from his shoulders. Jackson always knew what to say. And sometimes, what not to say. Their last deployment, after Rago got torn up in that roadside blast, after they'd kicked the shit out of the neighborhood and gotten their asses back to the FOB, after the adrenalin had drained from his veins and the shakes came on, Jackson had grabbed Pearson by the elbow and dragged him out of the tent and into the dark where he held him in a bear hug and let him cry it out. And afterwards, he never said a word about it.

Don't worry, Pearson says. *I'm not going soft.*

Back in the Shit

"What the fuck just happened?" Sergeant Payne screams, frantic, looking from the fire team of soldiers with their chest-slung M4s pointed at the ground to the three bodies lying on the small home's dining room floor. "This was supposed to be a snatch and grab."

On the sandstone floor is a man flat on his stomach atop a widening pool of red. Beside him in a heap by a hand-painted buffet is a teenage boy. On the other side of the table is a woman half-sitting in an overturned chair. Father, mother, son sprawled on the floor, their family dinner still hot on the table.

"Pearson and Shishlov came in," Corporal Parker explains, "told everyone to get down. But the kid made a move for an AK"—Parker holds up the captured weapon—"and my boys lit him up."

"And them?" Payne asks, his sweeping arm indicating the parents.

"The father could've been going for the gun too," says Parker, his face impassive.

The woman's body is far to the one side of the table, the two males at the other. From the doorway, Payne studies the angles and looks round at the men.

Pearson's gut draws together tight as a bear trap's spring, his jaw clenched, his brow bunched over his eyes. Beside him, Shishlov, Jackson's replacement, is wide-eyed and ghostly pale. For a moment, Pearson is certain Payne will ask the new guy for his version of events and who knows what Shislov will say. But then the heavy boots of second squad come tromping into the building, which means the L.T. and Platoon Sergeant Stackhouse won't be far behind.

Payne glances over his shoulder then back at the room. His face resolves like hardening clay. "Anyone search the bodies yet?" he asks.

Parker shakes his head.

"All right then. Get on it. Let's see if we can unfuck this cluster."

Shishlov, Pearson, and Willoughby spread the bodies on the floor, patting them down and emptying pockets, searching for weapons and intel. They find nothing, but the bodies are grouped together by the time Frisch and Stackhouse enter the room. Payne repeats the explanation, points out the captured AK-47, and the L.T. gets on the radio to relay the info up the line.

Pearson has his back to them, standing like a sentry over the bodies lined up on the rug like gutted fish laid out at market. The anger that had coiled inside his stomach for so long is gone now, leaving him cored out like a Halloween pumpkin. He stares at the slack faces, burning them into memory, not realizing how many times they'll visit his sleepless nights for years to come.

Behind him, Stackhouse grunts with his usual swagger. "Fucked. Them. Up," he says. "Hey, Pearson. Come here."

Pearson turns and shuffles over.

"Way to man up," Stackhouse says. He holds out his fist and Pearson stands frozen for a beat. Then Pearson bumps it with his own fist, his mouth flat and thin as the bladed edge of a bayonet.

Contributor Bios

Leonard Adreon served the US Navy from 1944 to 1946 and was a combat corpsman with the First Marine Division in 1951 and 1952. He is a graduate of Washington University in St. Louis. He is married and has three daughters, six grandchildren and one great-granddaughter. He facilitates writing classes at Washington University.

Jason Arment served in Operation Iraqi Freedom as a Machine Gunner in the USMC. He's earned an MFA in Creative Nonfiction from the Vermont College of Fine Arts. He lives in Denver, where he coordinates the Denver Veterans Writing Workshop with the Colorado Humanities and Lighthouse. He can be reached at jasonarment.com

Cheryl Ferguson Bernini is an American Expat living in Italy with her Italian husband and four felines. She grew up in a small town in Connecticut where her dad, Frederick M. Ferguson, was born and raised. He served in WWII as a member of the 280th Combat Engineers Battalion, Company A.

Michelle Brandfass is an aspiring author who has recently found a voice and an outlet for creative expression through writing by joining a program that is provided through the Department of Veteran's Affairs for disabled veterans. She is an Air Force Disabled Veteran married to a Recon/MARSOC Marine through most of his career.

Randy Brown was embedded with his former Iowa Army National Guard unit as a civilian journalist in Afghanistan, May-June 2011. A 20-year veteran with a previous overseas deployment, he subsequently authored the poetry collection *Welcome to FOB Haiku: War Poems from Inside the Wire* (Middle West Press, 2015). His poetry and non-fiction have appeared widely in literary print and on-line publications, including *Stone Canoe, Drunken Boat, F(r)iction*, and *So It Goes: The Literary Journal of the Kurt Vonnegut Museum and Library*. As "Charlie Sherpa," he blogs about military culture at www.redbullrising.com, and about military-themed writing at www.aimingcircle.com.

Nancy Brewka-Clark has a particular interest in writing short mystery fiction aided and abetted by her husband Tom, a crime reporter for many years. Her recent work appears in *Mysterical-E, Yellow Mama, Litbreak, Every Day Fiction, Close2theBone, Eastern Iowa Review, Malice Domestic's Murder Most Conventional*, and two short story collections and a book of poetry published by FunDead Publications of Salem, Massachusetts. Her short mystery plays have been produced here and abroad and published by Smith and Kraus, YouthPLAYS of Los Angeles, and Routledge U.K. The couple resides in Beverly, Massachusetts, where they met at the local newspaper after Tom's return from Viet Nam.

Steven Croft served with the 1st Battalion, 118th Field Artillery Regiment, 48th Infantry Brigade Combat Team, Georgia Army National Guard, from 2001-2012, with whom he deployed to Iraq in '05-'06 and Afghanistan in '09-'10. He is the author of two poetry chapbooks: *Coastal Scenes* (2002), and *Moment and Time* (2015), both published by The Saltmarsh Press. He works for the Brunswick-Glynn County Library in Brunswick, Georgia.

JD Duff grew up in the suburbs of New York City. She has a Master of Arts in Writing and a Master of Arts in Teaching English Education from Manhattanville College. Duff also holds an undergraduate degree in Africana Studies from Binghamton University. She taught college level writing and literature for over seven years. JD is married to a veteran of the Marine Corps and Navy and has also worked with many combat veterans to help improve their writings. Some of her publications may be forthcoming or found in *The Wrath-Bearing Tree*, *Storgy Magazine*, *Crack the Spine*, and *Melancholy Hyperbole*.

Michael Eaton served in the US Army from 1961-1964 in hospitals with an Occupational Therapy Technician MOS and completed service with a rank of E-5. He graduated from San Francisco State College with a Master's Degree in Creative Writing utilizing the GI Bill.

Jessica Evans is a Cincinnati native currently living abroad. She is a proud military spouse, married to an active-duty field artillery officer. Her work tackles serious themes and attempts to understand human motivation, duty, and honor. Work has appeared in various journals including *Gravel*, *Scissors & Spackle*, and the *Avalon Literary Review*. Her debut novel, *Hippie Mafia*, was published in 2017. When she's not screen-side, Evans is in the gym, Olympic weightlifting.

Emryse Geye is a graduate student, poet, and Army brat from the Pacific Northwest. His father is currently serving as a CW4 in the US Army. His work has been published in Tarleton State University's *Anthology*, the University of Northern Colorado's *Crucible*, as well as in Portland State University's *Pathos*.

Bill Glose is a former paratrooper and author of three poetry collections, including *Half a Man*, whose poetry arises from his experiences as a combat platoon leader in the Gulf War.

Shane Griffin is a graduate student at Iowa State University's Master of Fine Arts in Creative Writing and Environment. He is an Iraq War veteran and works as a firefighter/paramedic in Des Moines, Iowa. His work has appeared in the *After Happy Hour Review*, *Baltimore Review*, *Heroes' Voices*, *Collateral*, *Hippocampus Magazine*, *Sky Island Journal*, and the *Wapsipinicon Almanac*.

Andrew Gudgel is a retired US Army Warrant Officer and veteran of the Gulf War and Somalia. He's a freelance writer who lives in Maryland, in an apartment slowly being consumed by books. His poetry has appeared in Vol. 6 of *Proud To Be*, the St. John's College *Energeia*, and *Asimov's Science Fiction* magazine.

After eight years of active duty in the Air Force (including combat in Vietnam), then six years in the Missouri Air National Guard, **Jay Harden** completed a career in the Department of Defense, then became a photographer and writer of short stories, poems, and lyrics about love, war, childhood, and personal growth with award-winning work in seven anthologies. He also serves on the Veterans Issues Advisory Board of PBS station KETC

Ruth M. Hunt is a Master Sergeant in the US Army with 18 years of active duty service. She and her husband, James, are dual military, have been married over 15 years, and have three wonderful children. She is the proud daughter of a Vietnam veteran and has two sisters that also honorably served in the Army. She credits her success to the unwavering support of her family and overall trust in God.

Rob Jacques, a Vietnam-era naval officer, resides on a rural island in Washington State's Puget Sound, and his poetry appears in literary journals, including *Atlanta Review*, *Prairie Schooner*, *Amsterdam Quarterly*, *Poet Lore*, *The Healing Muse*, and *Assaracus*. A collection of his poems, *War Poet*, was published by Sibling Rivalry Press in 2017.

Monty Joynes has had three previous stories and two poems published in *Proud to Be*. His award-winning story "First Day at An Khe" appeared in Volume 1. Joynes is the librettist of a classical music oratorio, *The Awakening of Humanity*. He is the author of 22 books and continues to publish fiction, poetry, and non-fiction from his mountain home in North Carolina. Joynes served in the army (1964-1966) with the 91st Evacuation Hospital.

Patricia Joynes is primarily a nature photographer and has done cover shots for books, had photos in the official 2016-2018 Blue Ridge Parkway calendars, and had a photo selected in Oct. 2015 for *Nat Geo's* "Inside Access" story. Her husband St. Leger "Monty" Joynes served in the Army (1964-66) with the 91st Evacuation Hospital and won the fiction award in Proud to Be (Vol. 1).

Curt Last lives in Huntington Beach, California. He earned his Bachelor's Degree in Pre-Law from the University of California, Santa Barbara (1994) and his Master of Fine Arts in Poetry from California State University, Long Beach (2006). He served from 2008 to 2016 as a Hospital Corpsman in the United States Naval Reserves. Duties included various Navy clinics and hospitals, a humanitarian mission to East Timor (2010), and a deployment to the Role 3 Combat Hospital in Kandahar, Afghanistan (2011).

Attracted to words at an early age, **Rod Martinez's** first book was created in grade school. His teacher used it to encourage creativity in her students. His high school English teacher told him to try short story writing, he listened, and the rest, as they say, is history.

In 1967, **Ken McBride** served as a United States Marine in Vietnam. After being wounded in the battle of Duc Pho he was sent home where he finished his enlistment. He has written for the *St. Louis Post-Dispatch*, *Big Muddy*, *Reader's Digest*, *Proud to Be: Writing of American Warriors Vol 5*, *Massillon Independent*, and *Leatherneck Magazine*.

Carl "Papa" Palmer of Old Mill Road in Ridgeway, VA, now lives in University Place, WA. He is retired military, retired FAA, now just plain retired enjoying life as Papa to his grand descendants. Carl, Franciscan Hospice volunteer, is a former Pushcart Prize and Micro Award candidate.

Breanne M. Pye is a former U.S. Army photojournalist who is currently working on completing her MFA in creative writing at the University of Colorado at Boulder. Her Army photos and articles can be found in numerous news magazines, both online and in print.

Billie Holladay Skelley received her Bachelor's and Master's degrees from the University of Wisconsin-Madison. Now retired from the nursing profession, she enjoys focusing on her writing, and her work has appeared in various journals, magazines, and anthologies in print and online. An award-winning author, she also has written books for children and teens. Her father, Howard Kelly Holladay, served four and a half years in the Army Air Forces during WWII as a B-24 pilot.

Lauren Stevens is a member of the American Society of Journalists and Authors (ASJA) and makes her living writing for businesses on a freelance basis. She keeps her passion for writing alive by crafting creative nonfiction, much of which is about her experiences growing up in Cold War Europe as a military dependent. She has essays published in multiple anthologies, most recently receiving an Honorable Mention and publication in *Proud To Be: Writing By American Warriors, vol. 6*.

Ian D. Stochl is a full-time father, husband, and energetic nanoparticle chemist. He served in Iraq for the end of OIF 6 and all of OIF 7 as a M2 gunner on a security force mission with the army, although his original MOS was 74D. Using his education benefits he obtained a BS in chemistry from Saint Louis University. He is currently employed with NanoMetallix making an energetic nanocomposite.

Lauren Stochl is a dedicated wife to a combat veteran and mother to an amazing son. She was born and raised in Saint Louis, Missouri, which is where she currently resides. She is currently attending Saint Louis Community College where she is finishing her associate's degree and then intends to pursue a Bachelor's degree with a concentration in Biomedical Science.

Ryan Stovall is a former adventurer, world traveler, and Green Beret. His work has appeared or is forthcoming in *Rosebud, Geometry, The Cape Rock, Here Comes Everyone, The Deadly Writers Patrol*, and *As You Were: The Military Review*. Currently finishing his English degree, Ryan lives with his family near Bangor, Maine.

Bruce Sydow volunteered for combat missions as a Marine door gunner in Vietnam. Thirty-three members of his helicopter squadron, HML-167, were killed in action. Only 19 years old, he was one of the youngest Huey gunners in the unit to be decorated with the Combat Aircrew Wings, United States Air Medal, and Gallantry Cross. He graduated from college magna cum laude and holds a Master's degree from the University of Washington in Seattle. He was elected Professor of the Year, and was honored with the Excellence in College Teaching Award.

Major **A. Sean Taylor** enlisted with the Iowa Army National Guard on October 24, 2002, at the age of 35. He deployed to Bagram, Afghanistan with the Iowa Guard from 2010-2011 and to Taji, Iraq with the 310th ESC Advise and Assist Team in 2015 supporting the Iraqi Security Forces with their fight against ISIS/ISI.

Jonathan Tennis is a graduate of Eckerd College (BA) and Norwich University (MSIA). After serving in the US Army, he moved to Tampa, where he still resides, working as a consultant. He is currently pursuing his MFA in Creative Writing at the University of Tampa. His work has appeared in the *Eckerd Review, Military Experience and the Arts, O-Dark-Thirty, Odet, Proud to Be: Writing by American Warriors, vol. 6, Sanctuary Literary and Arts Journal*, and a festschrift in honor of the poet Peter Meinke.

Cynthia Teramae is an emerging writer based in Pittsburgh, Pa. She is a retired US Army Public Affairs Officer who served two tours of duty in Iraq. She recently graduated with honors from the University of Pittsburgh with a BA in Creative Writing.

Myrta Vida earned her MFA in Creative Writing from the University of Missouri –St. Louis. An award-winning writer, Ms. Vida currently works as an Assistant Editor for *december magazine*. Born and raised in Puerto Rico and a U.S. Army Veteran, Ms. Vida also teaches and leads creative writing workshops for a variety of Veteran- and minority-focused groups in Brooklyn, NY. She's an avid salsa dancer and brunch enthusiast.

Stacey Walker has taught college composition and literature for thirteen years in Southeast Missouri, but moved, with her husband and son, to Saint Louis in 2013.

There, she continues to teach and lecture at the college level at the University of Missouri-Saint Louis, Jefferson College, and Saint Louis Community College. Her husband, Kent Walker, is an Army infantry veteran who served two tours in Iraq in 2003 and 2007. Ms. Walker and her husband work with the Missouri Humanities Council to co-facilitate the Veteran's Writing Workshop. They are very proud to be a part of it.

Jeremy H. Warneke is a United States Army veteran, who served in Iraq. His publication credits include but are not limited to *Itscomplicated.vet*, *Fiction Southeast*, *NYC Veterans Alliance*, *Homefront Progressives*, *Task & Purpose*, and *Scintilla*. In 2017, he was a War Horse Writing Seminar fellow and a second-place poetry finalist for Line of Advance's COL Darron L. Wright Award. In 2015, he received an honorable mention for photography in *Proud to Be: Writing by American Warriors, Vol. 4*.

From 1966 to 1970, **Clayvon Ambrose Wesley** was an enlisted man in the United States Air Force. During that time, he spent eleven months and twenty-three days assigned to 483rd Hospital Group at Cam Ranh Bay Vietnam as a medic specializing in Intensive Care Surgical Recovery. At the time of discharge, Wesley had earned the rank of E-4 Sergeant. Currently, he is an artist with exhibitions on four continents.

While getting his MFA from the University of Tampa, **Ben White** thought he was a poet only to find that he isn't a poet at all. He is a witness. What he writes is testimony.

Charity Winters is a 2003 graduate of the United States Air Force Academy and a free-lance writer. During her six years on active duty, as an Air Force Security Forces officer, she completed three tours in Iraq. Her work has previously appeared in *Proud to Be* Volumes 2, 4, 5, and 6.

Valerie Elizabeth Young is a veteran of the United States Armed Forces. She served approximately ten years, with a deployment to Iraq and Hurricane Katrina. She is a Head Start advocate and parent ambassador. As a parent ambassador, she works with other Head Start parents to advocate for Head Start in the state of Illinois. Now also works with the Missouri Department of Corrections as a Probation and Parole Assistant.

Judge Bios

Ron A. Austin (Fiction) holds a MFA from the University of Missouri–Saint Louis, has served as an editor for *december* and *River Styx*, and is a 2016 Regional Arts Commission Fellow. His stories have been placed or are forthcoming in *Pleiades, Story Quarterly, Ninth Letter, Black Warrior Review, Midwestern Gothic, Juked*, and other journals. He has taught creative writing at the Pierre Laclede Honors College. He, his partner Jennie, and son Elijah live in St. Louis with a whippet named Carmen. His debut story collection, *Avery Colt is a Snake, a Thief, a Liar* won the Nilsen Prize and will be published in 2019.

Emma Bolden (Poetry) is the author of three full-length collections of poetry—*House Is An Enigma* (Southeast Missouri State University Press, 2018), *medi(t)ations* (Noctuary Press, 2016) and *Maleficae* (GenPop Books, 2013)—and four chapbooks. The recipient of a 2017 Creative Writing Fellowship from the NEA, her work has appeared in *The Best American Poetry, The Best Small Fictions*, and such journals as the *Mississippi Review, The Rumpus, StoryQuarterly, Prairie Schooner, New Madrid, TriQuarterly*, the *Indiana Review, Shenandoah*, the *Greensboro Review*, and *The Journal*. She currently serves as Associate Editor-in-Chief for Tupelo Quarterly.

Philip MacKenzie (Essays) earned an MFA from Minnesota State University, Mankato, and a PhD from the University of South Dakota. He is an instructor in the English Department at Southeast Missouri State University.

Missy Nieveen Phegley is the Director of Composition at Southeast Missouri State University. She has been published in *Class in the Composition Classroom: Pedagogy and the Working Class, English Journal, Kairos*, and *Cave Region Review*. She enjoys yoga, coffee, craft beer, and mountain biking, but she is way too old to be the hipster her interests suggest.

Seth Wade is an artist and educator in Dayton, Ohio. He received his BFA from the University of Dayton, where he is now an adjunct professor, and received his MFA from the University of Cincinnati.